EVOLUTION OF THE AMERICAN DIESEL LOCOMOTIVE

J. Parker Lamb

INDIANA UNIVERSITY PRESS
Bloomington and Indianapolis

This book is a publication of

Indiana University Press
601 North Morton Street
Bloomington, IN 47404-3797 USA

http://iupress.indiana.edu

Telephone orders   800-842-6796
Fax orders   812-855-7931
Orders by e-mail   iuporder@indiana.edu

© 2007 by J. Parker Lamb
All rights reserved

No part of this book may be reproduced or utilized in any
form or by any means, electronic or mechanical, including
photocopying and recording, or by any information storage
and retrieval system, without permission in writing from the
publisher. The Association of American University Presses'
Resolution on Permissions constitutes the only exception
to this prohibition.

The paper used in this publication meets the minimum
requirements of American National Standard for Information
Sciences—Permanence of Paper for Printed Library
Materials, ANSI Z39.48-1984.

Manufactured in the United States of America

Library of Congress Cataloging-in-Publication Data

Lamb, J. Parker.
Evolution of the American diesel locomotive / J. Parker Lamb.
p. cm. — (Railroads past and present)
Includes bibliographical references and index.
ISBN-13: 978-0-253-34863-0 (cloth : alk. paper)
ISBN-10: 0-253-34863-3 (cloth : alk. paper)
1. Diesel locomotives—United States—History.
2. Electro-diesel locomotives—United States—History.
I. Title.
TJ619.2.L36 2007
625.26'60973—dc22

2006032194

1 2 3 4 5   12 11 10 09 08 07

This book honors my close friends David Price and Tony Howe, whose patient tutoring allowed me to become proficient in digital rehabilitation of antique images.

CONTENTS

Preface
ix

1. Precursor Technologies
1

2. Self-Propelled Coaches
13

3. The Diesel Climbs Aboard
27

4. Streamlined Trains
39

5. Developments beyond La Grange
55

6. Alco Rebounds
67

7. Postwar Shakeout
77

8. Road Switchers Take Over
91

9. A Monopolized Market
107

10. Special-Purpose Designs
125

11. New Heights for Diesel Power
137

12. Recent Developments
159

13. The Diesel Century in Perspective
169

List of References
175

Index
177

PREFACE

Full-scale introduction of diesel-powered locomotives into the American railway industry occurred primarily during the fifteen years following World War II. Although this is a relatively short time for such a fundamental change, the actual gestation period of this revolutionary technology had taken much longer, as one would anticipate. Building on the highly documented segments of this significant sequence of events, the following narrative presents the evolutionary highlights of dieselization without delving into excessive details concerning each builder, each railroad, or each locomotive model. Moreover, the presentation not only includes technical aspects but also comments on the economic and social ramifications.

Clearly, these revolutionary machines evolved via the traditional trial-and-error melding of rudimentary concepts from disparate sources, followed by the absorption of periodic refinements of both electrical and mechanical elements, to produce a machine so powerful that it led to larger freight cars, stronger couplers, and more robust roadbeds. This enabled American railroads to continually set new world standards for Gross Ton-Miles per Train Hour, symbolic of the aggregate power produced by moving heavy trains at high speeds. Such results may well have been part of the motivation for the bold statement by a prominent rail commentator in March 2004 that the diesel revolution was "the greatest technological change of the twentieth century." While many would say that this is an exaggeration, others would agree that this was certainly one of the century's most important technical changes.

This book will follow the general approach used in my companion volume, *Perfecting the American Steam Locomotive* (Indiana University Press, 2003). Thus there will be emphasis on the role of leading engineers whose innovations paved the way for critical breakthroughs. Unlike steam locomotive evolution, which occurred in parallel with development of the nation's railway network, introduction of diesel power represented a revolutionary upheaval for an established sector of the economy that sent shock waves through rigid corporate cultures and staid government regulators. To some it represented the promise of enormous profits from rail operations, while to others it was an unmitigated disaster that personified the oft-repeated plight of business enterprises, namely, dealing with the introduction of a labor-saving technology that carried with it harsh consequences for many of its employees.

A recurrent theme in technological evolution is incrementalism, with progress on various components occurring simultaneously in many locations. Thus an understanding of these precursor (or enabling) technologies is extremely important. For diesel locomotives, there were separate developments early in the twentieth century on three key technical areas: a lightweight and efficient diesel engine, an effective electric-drive system, and a control system to modulate the power train. Only after these subsystems had reached a certain level of sophistication could the important process of integrating them into a commercially successful locomotive be undertaken.

The coverage will consider five developmental periods of the twentieth century.

Internal combustion powered coaches—These used mostly gasoline engines, but power to the wheels was usually by direct current electricity.

Early diesel-electric locomotives—Eight large companies built and marketed diesel locomotives before 1950, but only two had emerged by 1970 to advance this technology to a higher level.

Maturation period—The two major builders competed fiercely by offering continually improved designs for direct current locomotives, increasing power steadily to meet industry demands for higher speeds and longer trains.

Electrical revolution—Technical advances in solid-state electronics after 1960 permitted development of new devices that produced alternating current propulsion by 1993 as a more efficient alternative to traditional DC drive.

Manufacturing turnaround—In one of the greatest reversals in American industrial history, the last decade of the century saw an industry pioneer, and the leading diesel builder for over five decades, relegated to second place by an even older American company.

Proper coverage of such a broad subject requires the assistance of numerous colleagues and associates. I wish to express my deep appreciation to the following group of fellow rail enthusiasts. Especially useful were instructional manuals for early diesel units used by rail lines for their locomotive operators. These were supplied by Ed Mims, David Orr, and David Price, the first two of whom are former railroad executives. Reviewing early drafts of the manuscript to improve its accuracy and clarity was a time-consuming task that was carried out enthusiastically and successfully by James Mischke, a leading archivist of diesel history, and Mark Reutter, a widely published historian of American heavy industry. An extremely valuable contributor was my former colleague at the University of Texas, Mack Grady, an expert on power electronics, which represented the enabling technology for the realization of AC-drive locomotives.

To assist with illustrations, I was most fortunate to enlist some of America's leading photographers and photo collectors. From Arizona came historical Southern Pacific images from noted collector Arnold Menke, while Jim Shaughnessy of New York provided details of the *Flying Yankee* restoration and the Delaware & Hudson PA's. From the Midwest came the assistance of Kevin Keefe and the David P. Morgan Library of Kalmbach Publishing, as well as important additions from the vast collection of Louis Marre, who has archived and written about diesels for over four decades. A trio of my longtime friends from the Southeast, Jerry Lachaussee, David Price, and Louis Saillard, also provided key illustrations for this book as they did for the steam volume. In addition, Messrs. Mims and Mischke solved critical photo needs, while images of contemporary diesels came from fellow Texans Ted Ferkenhoff and Chris Palmieri along with Bill McCoy of Florida and Beth Krueger of Montana. A significant contribution was the artistic talent of Tony Howe, producer of the line drawings. As a computer expert, he assisted me and David Price with many of the digital renditions of antique images.

EVOLUTION OF THE AMERICAN DIESEL LOCOMOTIVE

Figure 1.1 The rapid transition from steam to diesel-electric power after World War II produced countless scenes like this one in Birmingham, Alabama. Antique 4-6-0 No. 7113, built by Baldwin in 1907 as Atlanta Birmingham & Atlantic No. 113, provides a stark contrast to purple-and-white GP7 No. 103. The colorful 1,500-hp Atlantic Coast Line diesel was less than a year old when photographed in May 1951, but the steamer would be scrapped within months. J. Parker Lamb.

CHAPTER 1

PRECURSOR TECHNOLOGIES

Internal Combustion Engines

The transformation of thermal energy into mechanical power using a piston-cylinder configuration was one of the fundamental concepts brought to practicality during the Industrial Revolution of the nineteenth century. In these machines, piston movement is produced by a working fluid that becomes highly energetic through a combustion process. In the intervening centuries, two major types of reciprocating heat power machines have seen widespread use. The earliest type was the steam engine in which the working fluid (water) receives its thermal energy outside the power cylinder (in a boiler). A half-century later, the level of technological development had progressed to the point where a second type of reciprocating engine was possible. In this design, the working fluid (air) undergoes an explosive combustion process inside the power cylinder. Hence the common names for these two designs are based on whether the combustion occurs inside or outside the cylinder.

The first internal combustion engine concept was proposed in 1862 by Alphonse-Eugène Beau de Rochas, who was later issued a French patent for a generic internal ignition engine, although no actual machine was ever built. Even at this early stage, theoretical calculations showed that such an engine would have a much higher thermal efficiency than a conventional steam engine. This efficiency represents the proportion of thermal energy in the working fluid that is converted to mechanical work.

Credit for building the first working machine belongs to Nikolaus August Otto and his partner, Eugen Langen of Cologne, Germany. Unaware of de Rochas's patent, they developed and demonstrated a single-cylinder spark-ignition engine at the Paris Exhibition in 1867. Otto and Langen then formed a company to produce these engines, and by 1877 they had received an American patent and begun marketing them in the United States. The scientific world has given Otto's name to the subsequent design of all spark-ignition machines (i.e., Otto cycle engines).

Another pioneer in improving the spark engine was Gottlieb Daimler, who was superintendent of Otto's engine works in Germany. In 1879 he patented a multi-cylinder machine with a crankshaft, and he soon teamed up with Wilhelm Maybach to produce the first commercially successful automotive engine. Another German inventor, Carl Benz, perfected a system for precise timing of ignition arcs from spark plugs. Of course, Daimler and Benz would later form a German automotive company whose successors are still in existence. Some estimate of the gestation period of engine technology can be made by observing that, at the beginning of the twentieth century, after over 52,000 steam locomotives had been produced in the United States, there were only 936 American automobiles using spark-ignition engines.

A second major type of IC engine was originated in the 1890s by French-born Rudolf Diesel, who had received a doctoral degree from the engineering school in Munich in 1880. After some years of working on ice-making machines, his attention turned to the elimination of electrically powered spark plugs in the Otto engine. Diesel's concept involved compressing the intake air to such a high pressure and temperature that, when droplets of fuel were sprayed into the cylinder, a spontaneous combustion would occur. His theoretical calculations demonstrated that the higher temperatures and pressures inside the cylinder would produce a thermal efficiency even larger than that of the Otto engine.

Early successes of the compression engine led to great wealth and acclaim for Diesel. For example, in 1897 St. Louis brewer Adolphus Busch traveled to Germany and negotiated the sale of American rights for the manufacture of his engine. The Busch-Sulzer Brothers Diesel Engine Company began production in 1910. Dr. Diesel made two trips to America in 1904 to visit the St. Louis Exposition, where he met with Busch and Edward H. Harriman of the Union Pacific. Some reports suggest that Diesel convinced Harriman to construct an experimental diesel locomotive. Although the trade press announced the testing, no further information was ever published. On his 1912 trip, Diesel met with Thomas A. Edison and other luminaries and gave an address to the American Society of Mechanical Engineers in which he chided Americans for being too timid in exploiting the advantages of his new engine concept. Had he lived a normal life span, he would have been more than vindicated, but the work of this eccentric genius was cut short in 1913 when, in poor health and despondent over recent business failures, he disappeared during a nighttime ferry crossing of the English Channel (a probable suicide). Subsequent development of the compression-ignition engine lagged far behind the Otto design because the former required a level of materials development and manufacturing precision that was still two decades in the future.

Because these two IC engine designs lie at the heart of diesel locomotive evolution, it is desirable to describe some thermodynamic features of each, using a pressure-volume diagram (P-V) to depict schematically the physical processes occurring inside the cylinder. In chart 1.1, which illustrates the operation of a spark-ignition engine, the top-most position of the piston is known as top dead center (TDC), and the bottom limit is the bottom dead center (BDC). The volume being measured on the horizontal scale is the space above the top of the piston at each instant. Each passage of the piston between top and bottom is known as a stroke. The directional arrows indicate two strokes in each direction. Thus this sequence of operations is known as a four-stroke cycle Otto engine and more commonly as a four-cycle Otto engine.

The diagram identifies six events:

1 to A   Intake stroke to pull air into the cylinder (intake valves open).
A to B   Compression stroke to raise pressure and temperature of fuel-laden air.
B to C   Spark ignition to further energize the fuel-air mixture.
C to D   Power stroke to rotate crankshaft.
D to A   Pressure decreases as exhaust valves open.
A to 1   Scavenge stroke to remove combustion gas from cylinder.

The four-stroke cycle requires two revolutions of the crankshaft for each power stroke. However, many engine designs are able to eliminate the first and last processes (represented by the dashed line in chart 1.1) and thus produce a two-stroke cycle engine. The applications of both designs to locomotive engines will be discussed later.

A similar P-V diagram of Diesel's original cycle is presented in chart 1.2a. The general

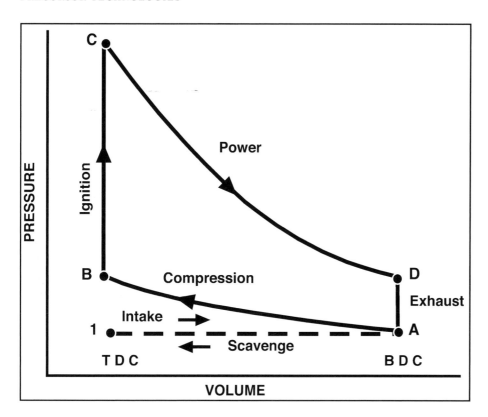

Chart 1.1. The four power-producing processes inside the cylinder of a spark-ignition engine (Otto cycle) are depicted schematically on a pressure-volume diagram. The pressure variation is that which exists at the top of the piston as it makes an upward stroke from bottom dead center to top dead center and return. The solid path characterizes a two-stroke cycle while the dashed line shows the four-stroke cycle. Howell and Buckius, *Engineering Thermodynamics*.

shapes of the two diagrams differ only between B–C, which depicts a constant pressure ignition. Also, it must be remembered that the maximum pressure in a diesel cycle can be at least two times higher than that in an Otto cycle. Moreover, the compression ratio (volume at BDC divided by that at TDC) is also much larger for the diesel cycle. While this diagram displays the operation of an early low-rpm diesel, one must represent a contemporary high-speed engine using some aspects of the Otto cycle, as shown in chart 1.2b, where it is seen that the ignition begins during a short period of constant volume near TDC. This is often denoted as a mixed cycle or mixed diesel cycle.

A major caveat in these P-V diagrams is that they represent idealized conditions. In particular, the sharp corners at the intersection of two processes would certainly not be realistic. Thus a measured pressure variation inside the cylinder would display rounded corners at these intersection points. However, the simple theoretical depictions are effective as indicators of engine characteristics.

Electrical Machinery

The basic principle by which an electric motor operates is magnetism, first discovered by the English physicist Michael Faraday in the 1820s. Although some materials are naturally and permanently magnetic, he showed that an electric current passing through a conducting material would also produce a surrounding electromagnetic field. An important property of magnets is their polarity (normally called north and south poles and noted as + or –). Magnets having the same polarity will repel one another whereas opposing polarities will attract. Soon after Faraday's discoveries, Joseph Henry, a college instructor in Albany, New York, developed a practical electromagnet and used it to power an experimental motor that employed attraction and repulsion to produce a reciprocating motion that was then transformed into rotating motion (just as in an early steam engine).

The first electric motors of significance were based on the current produced by a metal-

Chart 1.2. Pressure-volume diagram (at top) for a low-rpm compression-ignition engine (diesel cycle) features much higher peak pressures and temperatures than a spark-ignition engine with the same bore. These conditions produce ignition as the fuel is injected at point B. Both two-stroke and four-stroke cycle operations are indicated. At bottom is a corresponding diagram for a modern high-rpm diesel engine, in which ignition is initiated slightly before peak pressure. Howell and Buckius, *Engineering Thermodynamics*.

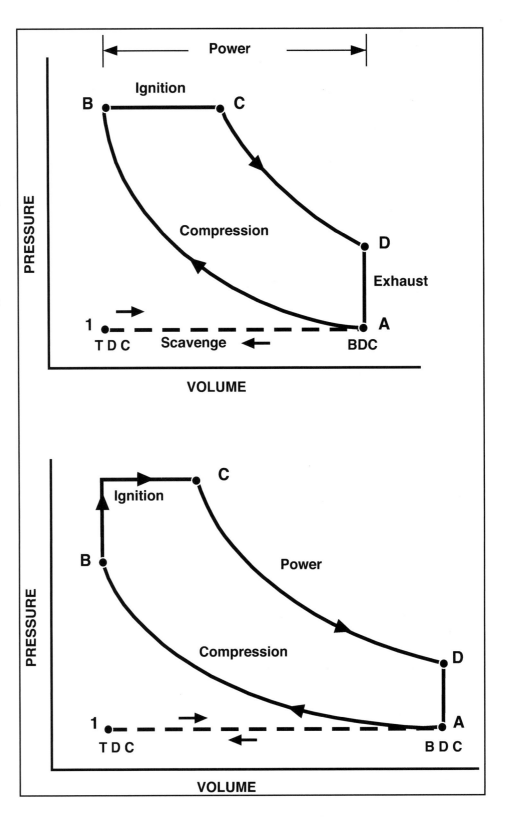

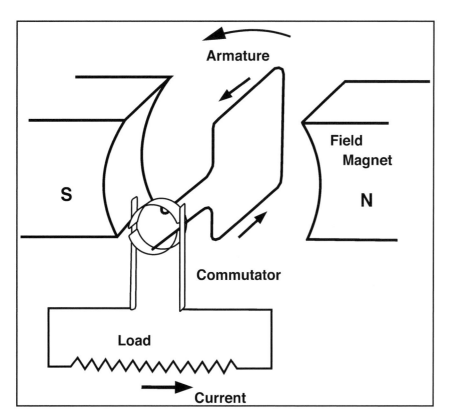

Chart 1.3. Fundamental elements of a two-pole direct current motor include the field magnet and the armature, which contains the commutator that is necessary to produce a unidirectional current flow.

acid battery, the first electrical storage device. Batteries operate through an electrochemical action that occurs when metal electrodes react with the surrounding acid. This results in a steady migration of negatively charged subatomic particles (electrons) from one side of the battery to the other, thus creating a voltage difference and consequent current flow between anode and cathode (+ and – electrodes). Since battery current is steady and unidirectional, it was designated as direct current (DC). Around 1840 Robert Davidson, an electrical inventor in Aberdeen, Scotland, completed a 5-ton locomotive powered by 40 zinc-iron sulfuric acid batteries, while the pinnacle of battery-powered trains came in 1851 when Dr. Charles G. Page constructed a 16-hp locomotive powered by 100 battery cells. The reciprocating action of the magnets drove a flywheel that allowed the train to reach 19 mph on a 5-mile line between Washington, D.C., and Bladensburg, Maryland.

The first practical DC motor was built in 1860 by Italian engineer Antonio Pacinotti utilizing an unusual ring armature. An even more important discovery was made in 1873 when Zénobe Théophile Gramme of Belgium demonstrated that the components of a DC motor and a DC generator were identical. In one case the machine received mechanical power (shaft) and produced a DC output whereas, if it received DC input, it produced mechanical power via a rotating shaft.

The sketch in chart 1.3 illustrates the rudimentary elements of a DC generator. Between the two magnetic poles (N and S) there exists an invisible electromagnetic force field. When a conducting wire passes through this field at a certain orientation, it is said to be cutting lines of magnetic flux (force). The result is an electric current in the conductor wire. By mounting the conducting loop on a rotating shaft (armature), one can transmit the generated current away from the machine using a commutator, illustrated by the split ring and the two electric brushes (rubbing contacts). Note that this is a simple two-pole machine with two segments in the commutator ring. Most traction motors have four poles, but other applications may have more. In addition, industrial-sized electromagnets generally consist of metal slabs wrapped by coils of wire known as field coils or field windings.

With the development of reliable DC motors, the era of electric locomotives began in

1879 when Dr. Ernst Werner von Siemens built a demonstration locomotive for the Berlin Exhibition. Running on a half-meter gauge track, the tiny train hauled 100,000 people during the exhibition. This demonstration mirrored a similar event almost 50 years earlier (1831) when Matthias Baldwin demonstrated his first steam locomotive with a miniature train operating inside a large museum building in Philadelphia. Three years after Siemens's demonstration in Berlin, Thomas Edison laid 1,400 feet of track at his Menlo Park, New Jersey, laboratories and developed a small DC locomotive that pulled two cars at 40 mph.

Within a few decades it became obvious that steam-powered commuter and elevated lines were producing widespread urban pollution, and there began a national effort to utilize these newly developed electric trains, which were both cleaner and quieter than smoke-belching steamers (even if they were tiny 0-4-2 Forney tank engines). Frank J. Sprague was one of the early engineering leaders of electric-powered railway equipment. In 1885 he developed a prototype for an electric-powered train using elevated trackage of the Manhattan Railway. Three years later, he constructed the Richmond Union Passenger Railway, the nation's first all-electric street rail system. He also developed the first front-hung traction motors mounted on a pivoted truck and geared to driving axle, and he later invented a method of coupling two electrically driven cars under one controller (the multiple unit connection).

Finally in 1894 General Electric's Lynn, Massachusetts, plant constructed the first power unit designed for regular railroad service, an experimental single-truck 30-ton model with traction motors on each axle. It was exhibited at the World's Colombian Exhibition in Chicago in 1896 and then put to work for an industrial railroad in New Haven, Connecticut. The next GE model was the nation's first steeple cab design, a 35-ton B-B unit that used four 500-volt DC traction motors to produce a drawbar pull of 14,000 pounds. It was eventually sold to an industrial mill and used until 1964!

As DC propulsion advanced, the proper method of controlling traction motors for varying speed and varying loads became important. The physical basis for motor control begins with the relationship between voltage, current, and resistance. This can be written as

$$\text{Voltage} = \text{Current} \times \text{Resistance}$$

where current is measured in amperes and resistance in ohms. This is known as Ohm's Law for circuits and can be applied to an entire circuit or to any element within the circuit. Symbolically this can be written as

$$E = I\,R$$

Moreover, electric power can be expressed as

$$\text{Power} = \text{Voltage} \times \text{Current}$$

or

$$P = E\,I = I^2\,R$$

Power is measured in volt-amperes, usually kilovolt-amperes (kVa). Power is also much more dependent on current levels than voltage.

The DC motor's two components, armature and field coil, have different electrical resistances. Moreover, there are two choices for connecting these components, as shown in chart 1.4. In a DC-series motor they are connected end to end, while in a DC-shunt motor they are connected in parallel. For a series motor, the total resistance of the two elements is merely the sum of the component resistances, and therefore, according to Ohm's Law, the voltage across each element will be different, although electrical continuity requires that the current through each element be the same. In contrast, with a parallel circuit, the full voltage acts across each component, thus producing different currents through each part, depending on its resistance.

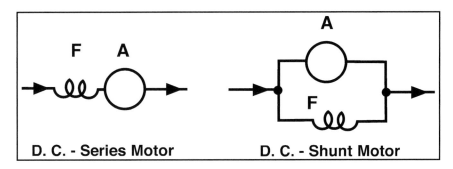

D. C. - Series Motor **D. C. - Shunt Motor**

Chart 1.4. The field and armature of a DC motor may be connected in either series or parallel arrangements, each of which has distinct operating characteristics. Middleton, *When the Steam Roads Electrified.*

Early experimentation showed that the DC-series arrangement was the most desirable for traction motor applications. As the load (current) on the motor increases, the magnetic field strength automatically increases, and the result is an increase in torque with a decrease in speed. Consequently, the series motor tends to adjust itself readily to the varying tractive effort required to accelerate and decelerate a train. In addition, further control can be obtained by grouping DC -series motors themselves in either series or parallel circuits or even in a hybrid series-parallel arrangement. These three circuits are depicted in chart 1.5, where four traction motors receive power from a generator. In one circuit, all motors are in series and therefore each operates on one-fourth of the total voltage of the generator. In the full parallel circuit, the total voltage of the generator is imposed on each motor, while the hybrid circuit allows each pair of motors to operate on half of the total voltage. For early locomotives, the series circuit was used for starting, with transition to the series-parallel at intermediate speeds and to the full parallel at high speeds. Moreover, the transition was a manual two-handed procedure. Later machines included a fully automatic transition control, while most modern DC machines use only two motor circuits, parallel-shunt and full parallel, where the former is a modification of the series-parallel connection in which part of the current is shunted from the field to the armature to increase power.

While General Electric surged ahead of everyone else in developing DC machinery after the 1880s, its major rival became engaged in a crusade led by its determined owner, George Westinghouse, who strongly favored the newer form of electricity, alternating current (AC). As its name implies, the current and voltage in an AC system change direction many times each second, and thus AC is characterized by an oscillation frequency measured in cycles per second (cps). One cps is denoted as one Hertz in contemporary literature, while one Megahertz equals one million cycles per second.

The creation of the first AC machines occurred in the 1880s. Lucien Gaulard and John D. Gibbs determined that one of AC's significant benefits was that it permitted a variation in voltage by use of a transformer, a device with no moving parts. This result led to the design of long-distance transmission lines for electric utilities and some electrified railroads. The remarkably simple transformer consists of two closely spaced iron cores (rods) that are wound with insulated wires. These windings are designated as primary (input) and secondary (output) coils. Early research showed that the difference in voltage between the input and output is determined solely by the number of turns (loops) on each coil.

For example, to increase the input voltage by ten times, the secondary coil (output) would have ten times as many turns as the primary. The basic principle involved is the same as mentioned earlier in connection with the simple DC motor, namely, a conductor cutting through magnetic lines of force. Thus the instantaneous voltage over one cycle of an AC signal rises from zero to a maximum value, then collapses to zero again before changing directions and repeating this up-and-down excursion. The number of increases and decreases in the primary voltage signal is multiplied by ten due to the larger number of secondary windings, thus producing a tenfold increase in output voltage. However, since no power is generated inside the transformer, we see that product of voltage and current is

Chart 1.5. Series-connected DC traction motors can receive power from the generator through series, parallel, or hybrid connections. Each arrangement provides performance characteristics suitable for different speed ranges. Middleton, *When the Steam Roads Electrified.*

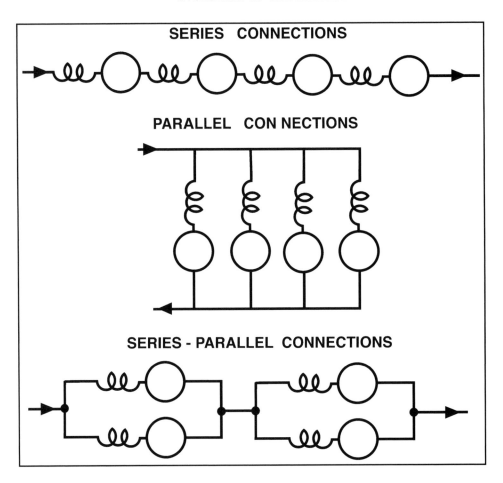

the same for both primary and secondary coils. Since voltage in the secondary is ten times larger than in the primary coils, the secondary current is ten times smaller. This is a significant advantage for long-distance transmission lines, inasmuch as the power loss equals the resistance of the line multiplied by the value of current (raised to the second power).

As an example, we consider a central generating station that produces a voltage of 1,000 (1 kV). A transformer then steps up the voltage to 10,000 volts (10 kV) and at the same time lowers the current by ten times before sending it out on the transmission line. Upon arrival at its destination, a step-down transformer converts it to a low voltage, high current signal that is useful locally.

George Westinghouse was a brilliant engineer who gained early recognition and wealth for his invention and development of pneumatic brakes for railroad rolling stock. He entered the electrical field soon after the Gaulard-Gibbs discoveries, and in 1885 he acquired U.S. patent rights for their designs. Only a year later, the Westinghouse Electric and Manufacturing Company completed the first commercial AC power-generating station in Buffalo, New York.

Recruiting a staff of stellar engineers that included the Croatian genius Nikola Tesla, Westinghouse Electric became the nation's leader in AC technology and a natural foe of Edison Electric Company (and its successor General Electric), which generally favored DC equipment. Eventually, both companies licensed each other's technology and were the dominant builders of electrical equipment. Westinghouse entered the railroad field in 1895, joining forces with Baldwin Locomotive Works to produce a small boxcab B-B unit (DC) weighing 46 tons.

As will be apparent in our discussion of diesel locomotive evolution, American manufacturing in the early twentieth century was dominated by two sectors, one with a mechanical orientation and the other emphasizing electrical machines and devices. The diesel-electric locomotive demanded expertise in both areas, and many existing companies were unable to develop such technical crossbreeding.

Locomotive Fundamentals

Following is some general background information that will assist in understanding subsequent discussions of locomotive development.

Engine configurations—The first IC engines had but one cylinder, but for commercial uses (e.g., automotive), at least two and preferably four cylinders were needed. Multiple cylinders required a long crankshaft that converted piston oscillations into rotating motion, in much the same way as a crankpin on the main driver of a steam locomotive. A slender rod, with a bearing at each end, connected the crankshaft to each piston. The lower ends of the connecting rods, as well as the crankshaft, were often designed to splash in the oil pan sump, which also included a pump to keep all bearings lubricated. At the top of the cylinders was a heavy cover (cylinder head) that sealed off the explosive combustion process, which occurred many times per minute in each cylinder.

Most IC engines use one of the following layouts of cylinders in relation to the crankshaft. The simplest one is an in-line configuration wherein all cylinders are aligned vertically above the crankshaft. The disadvantage of this layout is that, for more than eight cylinders, it produces long machines with slender crankshafts that are prone to transverse vibrations and fatigue cracking. Thus many designers began using a Vee-pattern in which the cylinders are mounted at a small angle (to vertical), thereby allowing them to be much closer longitudinally and producing shorter engines and crankshafts. While some early locomotive engines were inline models, most of those built after World War II are V-12s and V-16s.

An uncommon design for railroad applications but popular in other applications was the opposed-piston (O-P) arrangement, which appears to be two in-line engines stacked one above the other (head to head), so that there are two crankshafts, one at the bottom and another at the top. Therefore, each cylinder contains two pistons that oscillate in opposite directions. Major advantages of the O-P design are that it produces the maximum power in a given volume and automatically operates on a two-stroke cycle. It needs no valve control mechanism, since one piston serves as the inlet and outlet valve for the other. A fourth engine layout, used only in cases of limited space and small power requirements, is known as a pancake configuration wherein the cylinders are lined up horizontally on each side of the crankshaft. Schematically, it is the configuration that results if the angle between cylinders of a Vee-layout is increased to 180 degrees. Such a design has been used often in small automobiles, airplanes, and boats.

Engine performance—The following discussions will often include only two pieces of information about the output of an IC engine, horsepower and rpm, even though there are four critical measures of engine output. These are compression ratio (defined in relation to chart 1.2), cylinder displacement (volume displaced by piston head during one stroke), rpm (number of power strokes per second or minute), and finally the form of the aspiration process (how incoming air enters the cylinder). Greater engine power can be obtained by increasing compression ratio, displacement, and rpm. Additional power is also possible by providing augmented aspiration so as to increase the density of air entering the cylinder. With such heavier air, the piston will be able to do more work on each power stroke and hence output is increased.

A simple theoretical analysis identifies an important figure of merit for engine performance. Known as the Mean Effective Pressure (MEP), it is proportional to the compression ratio and the density of incoming air (aspiration), and it is related to engine output in

Chart 1.6. Variation of
locomotive tractive effort
(pulling force) with speed
for two types of DC units,
illustrating the effects of
adhesion limits and
additional motors. Marre
and Pinkepank, *The
Contemporary Diesel
Spotter's Guide.*

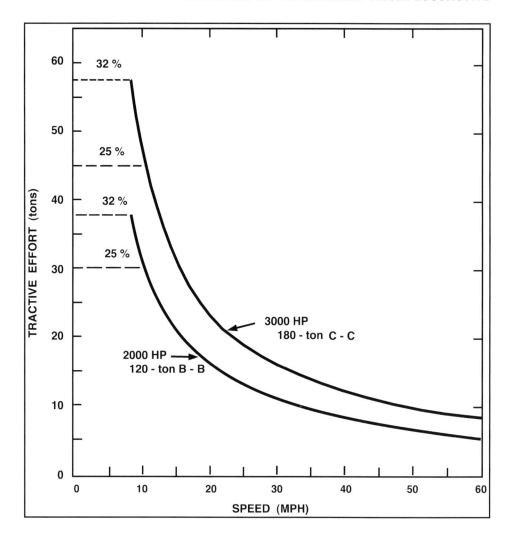

the following way. The mechanical work done during one stroke of the piston can be expressed as the product of the length of the stroke and a hypothetical mean effective force (MEF) that exits during the stroke. However, the MEF equals the MEP times the area of the piston head, so that mechanical work is

$$\text{Work (per stroke)} = \text{MEP} \times [\text{Area} \times \text{Stroke}]$$

The terms in brackets are equal to the engine displacement, a fixed value. This means that the power produced by one piston equals the work (per stroke) multiplied by the number of power strokes per minute. This latter number is proportional to the rpm and the type of engine (two-stroke or four-stroke cycle). Therefore, engine power is proportional to rpm, engine displacement, and MEP, with the latter being a function of compression ratio and aspiration. Engineering literature usually employs the term *Brake Mean Effective Pressure* (BMEP) to denote the value of MEP computed from actual performance tests.

Only a four-stroke cycle engine can operate solely with the suction of outside air during the downward (intake) stroke. The simplest method of augmentation is to push more air into the cylinder with a blower (high-pressure fan) that provides a small level of pressure increase above atmospheric. Most early locomotive engines used this technique. The largest rate of augmentation is provided by a supercharger (compressor), which can be either gear-driven or powered by a turbine operating from hot exhaust gas. It is also commonly used in locomotive applications. The supercharging process has virtually no effect on the shape of the P-V diagrams in chart 1.2. However, the entire cycle operates at larger pressures than does a normally aspirated engine. As a historical footnote, super-charged aircraft engines saw major development during World War II in order for bombers to fly at higher altitudes (where the air density is low) and thus were less vulnerable to ground-based artillery.

Locomotive Performance—Identification of the running gear of electrically powered railroad trucks, irrespective of the source of power, is based on a European system that counts axles, unlike the Whyte system for steam locomotives, which counts wheels. Powered axles are identified by letters, whereas unpowered ones (called idlers) are numbered. The letters A, B, C, and D represent one to four axles, respectively. The most common power design, a two-axle, two-motor configuration, is a B-truck. In early motor cars, only the forward truck was powered, thus its running gear was designated as B-2. Most locomotives rode a pair of similar trucks, B-B or C-C, where the dash indicates a connecting locomotive frame. In other cases, notably for passenger locomotives, a three-axle power truck was used for a smoother ride at high speed, although the light tonnage required only two powered axles. This led to the A1A–A1A configuration. While the foregoing arrangements were used most often during the early years of diesel evolution, many other combinations were employed occasionally. These included B-1, B-2, B-A1A, D-D, and B+B–B+B where the plus sign indicates the locomotive main frame riding on two pair of B-trucks, each pair being connected with a bolster.

The importance of the number of powered axles is shown in chart 1.6, which presents the variation of tractive effort with locomotive speed. Both four-motor and six-motor locomotives (B and C trucks) are included, along with two locomotive weights. It is also evident that, while the additional motors of the C-trucks provide a substantially higher tractive effort below 20 mph, they provide much less augmentation above 40 mph. However, the larger locomotives could pull more at low speed and thus accelerate faster. Because of their greater versatility, they have been more popular than B-B units on some railroads.

Another important aspect of performance from chart 1.6 is the adhesion values, which determine the maximum tractive effort as a percentage of the total weight on the driving axles. If the locomotive attempts to pull more load than is allowed by the adhesion limit, the wheels will slip and suddenly lose all power. The usual adhesion limit for dry rail is 0.25 as indicated by the 30-ton pull for a 120-ton locomotive. By sanding the rails, this value can be improved to 30 percent. Since the largest tractive effort occurs just before wheel slip, modern locomotives include wheel-slip control systems that use sensitive electronic measurements to detect any small difference between the motion of the ground and that of the wheel rim. For normal operation, both of these must be the same. But as the wheel begins to slip, it speeds up instantly. By detecting an incipient slipping condition with highly accurate measurements, the wheel-slip control system slows the wheel speed by short interruptions of current to the motors. These systems permit overall adhesion values over the 30 percent maximum shown. The latest truck designs have resulted in even further advances in adhesion (see chapter 11).

Figure 2.1. Oregon & California McKeen No. 41 is about ready to depart Salem, Oregon, in 1915. This 55-foot, 30-ton car was constructed in 1909 with builder's number 62. It was transferred to O&C's parent company, Southern Pacific, in January 1926 and scrapped seven months later. H. L. Arey, Arnold Menke Collection.

CHAPTER 2

SELF-PROPELLED COACHES

At the turn of the twentieth century, railroads were beginning to examine ways of reducing costs of passenger trains that served lightly used branch lines, sometimes in competition with electrified interurban railways. Clearly a small steam locomotive pulling one or two cars represented an exorbitant waste of power as well as money for fuel, equipment maintenance, and crews. It was obvious that a less expensive but equally effective alternative was needed. Earlier attempts had been made to develop a powered coach for use on low-income runs using small steam engines, storage battery-powered electric motors, internal combustion engines, and even compressed air engines. The primary barrier was to determine the best type of transmission, and the choice was soon narrowed to mechanical or electrical drive.

McKeen Motor Car Company

A major railroad figure stepped into this scenario in 1904 when Edward H. Harriman, who controlled both Union Pacific and Southern Pacific, summoned his superintendent of Motive Power to UP's New York office. Indiana native William R. McKeen, an egotistical and flamboyant character, was a brilliant mechanical engineer and no-nonsense manager who had become UP's power chief in 1902. Harriman's instructions were simple and direct: assemble a team in Omaha and produce a rail car powered by an internal combustion engine. Although the UP shop forces knew how to make a car body, they had absolutely no experience with gasoline engines. Thus the team decided to buy an existing engine and install it in a special car body. McKeen himself sketched the exterior of the first model (designated M-1), a 30-foot, single-truck car whose distinctive appearance featured a shiplike pointed nose and large steel frame known in the steam era as a cowcatcher.

The new design incorporated some visionary features, such as its low floor-height for easy platform access and a strong body structure that tied the underframe, sides, and roof into a tubular structure that would later prove to be extremely crashworthy. The company that produced the first McKeen design was Hale & Kilburn of Philadelphia. As it happened, their shop superintendent was a young engineer named Edward G. Budd, who would later make major passenger train innovations through his own company. McKeen's group searched for a large and rugged gasoline engine and finally selected a six-cylinder 100-hp marine motor constructed by the Standard Motor Construction Company of New

Figure 2.2. The engineer of Union Pacific M-25 seems to be ready to load his train at Hines, Oregon, in July 1938. This 70-foot, 41-ton car was built in 1910 as Santa Fe M-102 and sold to UP in 1922. It was not scrapped until 1944, making it one of the longest-operating McKeens. Note the large driving wheel below the engineer as well as the flywheel of the crosswise-mounted engine. Robert Searle Collection, courtesy of Arnold Menke.

Figure 2.3. Rearward view of Southern Pacific No. 57 at the Sacramento shops in 1934 shows the characteristic McKeen porthole windows and the tubelike car body construction with door at platform level. Guy L. Dunscomb, Arnold Menke Collection.

Jersey. It was the largest power plant that they could fit into the allotted space. However, the manual transmission they devised was something of a mechanical nightmare, with a hand-operated friction clutch, sliding gears, and a chain-driven driver axle.

Not surprisingly, the M-1 required considerable skill for its operator to accelerate and decelerate between stops. Starting the engine was also an adventure, since there was no battery power aboard. With three cylinders open to the atmosphere, compressed air was used to drive the other three pistons until the rpm was high enough for the fuel to be admitted to the originally closed cylinders. Once the engine was running, all cylinders were switched to fuel. To change gears, the operator closed the throttle to let the engine idle, then synchronized the gearing and slid the transmission into a different gear ratio. All of this was done with levers that were positioned at the operator's side rather than in front of him.

The first car, its sides resplendent in glossy maroon paint, rolled out of the shop in March 1905 and began a month of test runs around Omaha. In a test of its pulling power, it easily lugged a standard mail car up a 1.6 percent grade on the shop's coaling trestle. It was then assigned to service out of Portland, Oregon, where it would be required to operate up a 4 percent grade leaving the downtown area. Crews noted that when the M-1 was slowed to a crawl on the steep grade, it never stalled out, which was fortunate, since a portion of the grade traversed the city's red-light district.

After its Portland tests, the M-1 began regular service between Kearney and Calloway, Nebraska. Although there would be numerous operating problems with the car's new gasoline engine and complex transmission, its initial novelty created widespread sensation, a reaction that would be repeated 30 years later with the early streamliners. The Omaha *Illustrated Bee* suggested that "the gas car will supplant the locomotive." McKeen boldly predicted that "future use of these cars will be enormous." It was in Kearney yard that switchmen reportedly began referring to the midget car as the "potato bug." The bug title was only the first of many nicknames that would be applied to motorized coaches over the next 40 years.

Anxious to move forward on their initial success, McKeen's men produced the M-2 in September 1905. This 55-foot-long, 55,000-pound, all-steel car, called the battleship, was more conventional in design, with two trucks and a capacity of 57. Its body was half as heavy as a conventional coach but 25 percent stronger due to the tubular framework. It was also improved mechanically with an air-operated clutch and a McKeen-designed rear truck that produced an especially smooth ride. Soon the Omaha works turned out cars M-3 through M-6 and scattered them around the western two-thirds of the nation on demonstration runs, including the Alton Railroad in Illinois, Southern Pacific in Texas, UP's Oregon Railway and Navigation subsidiary at Portland, and Pacific Electric in Los Angeles. Eventually the M-6 replaced a four-car steam-powered train between Leavenworth and Lawrence, Kansas, after a short four-wheel trailer was added for baggage and express.

The M-7 model, produced in 1907, was the first to include what would become distinguishing features of later McKeen cars (besides the sharp prow), a low center-door (obviating the use of a porter's step), and shiplike porthole windows 22 inches in diameter. The 70-foot car also contained space for baggage and mail so that a trailer car was not required. Unfortunately, M-7 also highlighted the most undesirable feature of McKeen's cars: the noise and vibration produced by the opening in the forward floor that allowed the truck-mounted engine to swivel. The following M-8 model car displayed a number of advances, most notably a new 200-hp engine designed by McKeen himself. Its new carburetor was much more economical, allowing the engine to operate on level track using only one-third of a gallon per mile. It also employed an improved control system that indicated to the operator exactly when to shift gears.

In late 1907, after M-7 had demonstrated on the Chicago & North Western and Erie while en route to a New York City reception hosted by Harriman, the chairman decided it was time to spin off motor car construction as a separate business unit, with McKeen as president. Thus the McKeen Motor Car Company was incorporated in Nebraska on

Figure 2.4. One of the last McKeens in service was Virginia & Truckee No. 22, shown at Carson City, Nevada, in 1947. This car is now being restored by the Nevada State Railroad Museum. R. H. McFarland, Arnold Menke Collection.

August 3, 1908, and immediately established operations in a leased building on the north side of the UP shop complex, staffed with 50 employees.

Around 1913 McKeen decided to shift to a larger, more powerful car in order to compete with standard trains by pulling a 50-foot trailer car that also had the characteristic porthole windows. The new design contained a 300-hp engine having a much larger cylinder diameter (bore), while the trailer also incorporated a revolutionary idea: ball bearings in lieu of friction journals. The Hyatt Company, whose president was a young Alfred P. Sloan (later of General Motors), supplied these new components, the first to be used on railroad rolling stock. Another McKeen innovation was the use of an alternative fuel. Some cars were fitted with modified carburetors that allowed the engine to start by using gasoline but then switch to a cheaper and less volatile fuel known as distillate. The period around 1913 represented the zenith of McKeen car usage, with nearly 140 in service nationwide.

As World War I loomed, however, it became obvious to all but McKeen himself that his unreliable and labor-intensive mechanical drives had been supplanted by the electric motor. Thus in 1920, with zero sales and mounting debt, the McKeen company was purchased by Union Pacific and closed. But during its 15-year life span, McKeen's venture had proved that powered coaches could play an important role in early-twentieth-century railroading and had shown the internal combustion engine to be a practical alternative to steam power.

GE Gas Engine Division

Parallel to the development and usage of mechanical-drive cars in the United States, a gasoline-electric car had been constructed in England and operated with limited success. Industry giant General Electric was quick to notice this potential new market for its DC generators and traction motors, and in 1904 GE organized a gas-electric project group within its Railway Engineering Department at Schenectady. The three lead engineers were Henri Chatain, A. F. Batchelder, and S. D. Priest. The first two men initially surveyed the potential gasoline engines available in the United States and found nothing that would fit their needs. They finally turned to Woolsey Tool & Motor Car Company, the

English automobile company whose engine had powered the earlier gas-electric car, and placed an order for a six-cylinder, opposed-piston 9 × 10 (bore-stroke) engine that produced 140 hp at 450 rpm.

While the engine was under construction, GE negotiated with the nearby Delaware & Hudson Railroad to supply a car, and they offered a truss rod–supported Smith & Barney wooden combine. After GE's remodeling, the baggage area held the engine and a second control station was installed in the rear. Power to the axles was supplied by two streetcar-type traction motors (75 hp each), which were driven by a 120-kW, 600-volt generator. With great anticipation, the engine arrived for installation in 1905. Upon opening the 7-ton crate and examining the contents, GE engineers realized, as had their McKeen contemporaries, that these early engines were extremely complex machines and quite difficult to operate. Besides that, the engine was far too heavy and caused the car to weigh 68 tons. On its first trip in February 1906, D&H 1000 ran smoothly but could go no faster than 40 mph (with a 35 mph average speed). The car ran off 5,000 miles in service on the D&H, but eventually it was purchased by GE and rebuilt for sale.

Chatain and his group learned many lessons from their first machine, the principal one being that General Electric would have to construct its own engine. Consequently a new Gas Engine Department was established in 1906, and its staff soon settled on a V-8 configuration producing 100 hp at 550 rpm (a high speed for the time). Most important, the new GM-10A1 weighed slightly less than 2 tons in contrast to nearly 7 tons for the Woolsey design. For the second car, a typical streetcar controller was used, with a manual transition between series and parallel connections for the generator and traction motors. The operator's control stand also contained necessary engine controls: throttle, spark timing, and reverse. The 50-foot steel shell for No. 2 was constructed by Wason Manufacturing Company of Springfield, Massachusetts. Seating 44 (plus a 6-foot baggage area), the arch-roofed car resembled a contemporary interurban car.

Total weight savings for the second car was over half from that of the D&H car (31 tons compared with 68). Of course, the lighter car produced much higher acceleration (1 mph per second up to 25 mph) and a top speed of 55 mph. The first public demonstration of No. 2, again on D&H trackage, occurred on January 15, 1908. It then left on an extended midwestern tour, during which time it was damaged in a head-on collision with a 4-6-0 in South Dakota. Rebuilt as GE No. 7, it was sold to Minnesota's Dan Patch Lines in December 1910, but it was destroyed by fire in 1914.

The performance of GE's V-8 in No. 2 encouraged Chatain's group to design an improved version (GM-10B1) that had slightly larger diameter cylinders and produced 125 hp. It also featured a new front truck design powered by two 100-hp motors, while avoiding the previous practice of using a black powder charge to start the engine. Instead, compressed air (from a 4-hp auxiliary engine-compressor) was used to initiate rotation, while magnetos operated the spark plugs. With the delivery of Car No. 3 in May 1909, it was clear that GE had been influenced by the McKeen car's pronounced prow, inasmuch as No. 3 displayed a more parabolic nose than had No. 2. In an effort to cast a wide marketing net, GE offered a dual-power version of car No. 3 that included trolley poles for straight electrical use. Not surprisingly, no such cars were ever built. However, the extended demonstration tour had generated some orders, and General Electric launched its line of production cars in 1910.

The main feature of the production cars was another new engine (GM-16A1) in which the included angle between the two banks of cylinders was reduced from 90 to 45 degrees and the single crankshaft was replaced with one for each cylinder bank. The burning of exhaust valves (due to high combustion temperatures) was eliminated by installing vent ports at the bottom of the stroke. The new power plant also featured two carburetors, one for low rpm (200 or less) and the other for higher speeds. With the motor car business booming, GE moved it to the new Erie, Pennsylvania, plant before delivery of the first

Figure 2.5. General Electric teamed with Wason Manufacturing Company to produce large gas-electric cars that featured a curved prow similar to that of the McKeens. A glossy black Illinois Central No. 113 is being serviced at Vicksburg, Mississippi, in 1924. This 54-ton car, built in November 1914, carried 84 seats and was powered by a 175-hp engine. It worked for IC until 1942. C. W. Witbeck, David Price Collection.

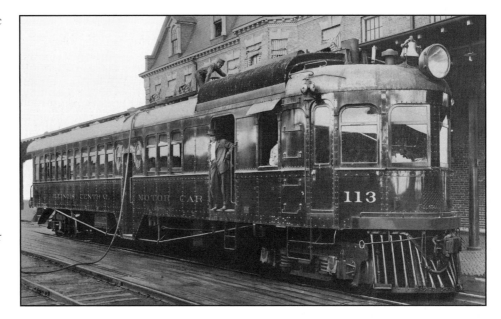

production car in February 1911, with 16 more to follow that year. During the following three years, the company produced 61 cars, representing the bulk of its total output, since only 8 cars were sold between 1915 and 1917.

As the demand for gas-electric passenger cars slackened, GE began building locomotives, short platforms with two power packages and double-ended controls. The first of these was Dan Patch Lines' No. 100, shipped from Erie in July 1913, with three more following in 1915. But, as with the McKeen Company, the onset of World War I signaled the end of GE's gas engine power program. The company reentered the railcar business in 1923 as the primary supplier of the electrical gear required by car builders such as J. G. Brill, Electro-Motive, Hall-Scott Engine Company, and American Car and Foundry (ACF), which later purchased Brill and Hall-Scott.

Although GE's output of gas-electric cars encompassed only 94 units, there would be two long-lasting effects. The first was that some members of the engineering team, notably Richard Dillworth and James Heseltine, used their valuable experience at GE in later developments of gasoline- and diesel-powered cars. In addition, the work of a senior GE electrical engineer, Swiss-born Hermann Lemp, would also have a profound effect on later developments. As early as 1910, Lemp had become a consultant to the Gas Engine Department, and he immediately began applying his substantial intellect to the vexing problem of controlling power systems consisting of an IC engine running an electrical generator. Not only would Lemp become one of the pioneers of the field known as automatic control but his meeting with Rudolf Diesel during Diesel's visit to Schenectady in 1910 caused him to become a strong advocate of the compression ignition engine for railroad use. This part of the General Electric story will continue in chapter 3.

J. G. Brill Company

Both gas-mechanical and gas-electric cars were produced by the Brill Company, which was founded in 1868 by Johann Georg Brill, who had immigrated to Philadelphia from Germany in 1847. Beginning with construction of horse-drawn streetcars in the1870s, the company built steam-powered streetcars and cable cars before concentrating on electric trolley cars and trolley buses. Its early specialty was car construction, and thus it bought

Figure 2.6. Brill's smallest gas-mechanical car was the 42-foot Model 55 powered by a 68-hp Service Motors engine. Number 500 was built in 1923 for the 92-mile Tennessee Alabama & Georgia, which connected Chattanooga with Gadsden, Alabama. In a June 1951 scene at Gadsden, the *Scooter*'s crew was waiting to get the highball for return. J. Parker Lamb.

Figure 2.7. Baltimore & Ohio No. 6040 was a 60-foot Brill Model 250 (denoting its horsepower) built in January 1927. This scene is in Parkersburg, West Virginia, in August 1946. A year later, the car's Brill-West engine would be replaced by a Cummins diesel, adding 7 tons to its original 48-ton weight. James Mischke Collection.

trucks and power sources from Sprague Electric, GE, and others, but eventually Brill developed its own truck designs.

In the 1920s, Brill began building gas-mechanical cars using engines from Mack-owned International Motors Company with later models using four-cylinder designs from Service Motors Company, which it later bought. In 1922 it began offering a 42-foot Model 55 that included both passenger and baggage compartments. Later designs in 1924 included Models 65 and 75, which used six-cylinder engines in order to pull trailer cars. Model 75 was 55 feet long with a top speed of 60 mph. Many of these Brill cars were built for overseas customers, including in Australia and Cuba.

In 1925, the company moved into the gas-electric car market and became a competitor for EMC until the Great Depression. Major railroads using Brill gas-electrics included AT&SF, B&O, Lehigh Valley, New Haven, Pennsy, Reading, SP, and UP. Their general appearance was more like the rounded GE and McKeen cars than it was the box shape of EMC car bodies. The company's production run of 184 cars ended in 1932.

Electro-Motive Corporation

The period between 1917 and 1920 was not a good time for American railroad innovation. Although governmental control by the U.S. Railroad Administration led to the introduction of rugged, standardized designs for locomotives and other rolling stock, neither locomotive builders nor railroad presidents were allowed to use their discretionary budgets for R&D work on new propulsion systems. Thus it was not until 1922 that another savvy visionary stepped onto the winding trail of diesel locomotive evolution.

Harold Lee Hamilton, a Californian born in 1890, was a Southern Pacific call boy at age 15 and held a number of railroad jobs before contracting malaria in 1911 while working on the Florida East Coast. While recuperating in Denver, he took a job in the sales and service department of White Motors, builder of large trucks and buses. After a short stint in military service during World War I, he returned to White Motors and was assigned as manager of the Des Moines, Iowa, district office. During this period he became quite familiar with the line of converted highway buses that White marketed to railroads after replacing the usual running gear with flanged wheels. As these vehicles were not constructed to the rugged standards of railway equipment, they were highly unreliable due to constant breakdowns. Consequently, Hamilton voiced negative comments to the company regarding any future marketing of these vehicles, but his advice was ignored. However, the persistent Hamilton did press his point to the extent that White Motors did not attempt to sell any more rail buses in the Des Moines district.

With these experiences bringing him in close contact with rail operations, he began studying the recent experiences of Brill, GE, McKeen, and other builders of IC engine–powered cars. Soon he became convinced that the electric drive, pioneered by GE, represented the best route for future development. This decision led him to take a leave from White in 1920 and explore opportunities for advancing the gas-electric coach beyond the prewar GE designs. He was sufficiently confident of a positive outcome that he resigned from White in July 1922 and began preparing a business plan for a new car company. A month later, he was one of the incorporators of the Electro-Motive Engineering Corporation of Cleveland. He then began to tackle the next challenge for a start-up company, namely, locating individual or corporate investors who would underwrite his plan of development.

One by one, prospective investors declined to invest in Hamilton's unorthodox idea of running Electro-Motive as an equipment consolidator. With no assembly facilities of its own, EMEC would purchase all drive-train components from vendor companies and hire an established car builder to assemble them. However, to ensure quality workmanship, all assembly would be carried out under the supervision of EMEC engineering

Figure 2.8. EMC–St. Louis Car gas-electric No. 2500 of the GM&O is waiting for the highball at Meridian, Mississippi, for a run to Jackson, Tennessee, in 1950. Built in 1927 as No. 1821 for the Mobile & Ohio, this 73-foot, 64-ton car was powered by two Winton 106-hp engines. The twin engines account for extra large radiators that occupy most of the car's front face. W. H. B. Jones, David Price Collection.

managers. General Electric was quite willing to provide the electrical package, but after its rocky experiences with building large engines, it wanted no part of this component. Many observers have suggested that GE, whose corporate traditions and culture were skewed toward electrical devices and machines, was never comfortable having a Gas Engine Department. Of course, these attitudes would change, but it would take more than two decades.

In his search for an engine builder, Hamilton soon found himself at the offices of nearby Winton Engine Company, a privately held firm formed in 1912 by Alexander Winton, cofounder of the Winton Automobile Company of Cleveland. Winton's engine plant was a successful producer of marine engines originated by both Otto and Diesel. The response from Winton engineers about building a new engine for railroad use was positive. But, as often happens, there was a major barrier. Most of the owner's assets were unavailable due to the receivership of the automobile company. Consequently, the official response was "No, we do not have available capital to invest in such a speculative venture."

But the salesman inside of Harold Hamilton did not take this as the final answer. Soon he and George Codrington, general manager of Winton Engine, visited the home of Alexander Winton to discuss the new engine needed by EMEC. As Hamilton related at the 1955 U.S. Senate antitrust hearings, "As we discussed our plans with him, a transportation pioneer now in his last years, I saw his eyes light up at the thought of doing something novel in the field of railroading. After Codrington summed up the plan, Winton rose from his chair and, laying his hand on the General Manager's shoulder, said, 'Go ahead and build it, George, and send me the bill.' " With this decision, Winton's company had just sealed its destiny as a pioneer in the American diesel revolution.

The year 1923 was a pivotal one for Hamilton. First, he removed the word *Engineering* from the company name to discourage outside requests for general engineering work, in which he had no interest. Then, after having persuaded Edwin E. Meissner, president of St. Louis Car Company, to invest in his company and carry out the final assembly of EMC motor cars, construction of the first unit began in October. It was a 35-ton, 57-foot, 44-seat car

Figure 2.9. On an August day in 1950, the platform crew at Lubbock, Texas, hurries to finish their chores so that Santa Fe M-182 can move out. The 80-foot, 70-ton product of EMC-Pullman (1929) carried a 400-hp Winton Model 148 engine, which was replaced with a diesel in 1946. The car was not scrapped until 1958. J. Parker Lamb.

for the Chicago Great Western. Its six-cylinder, 175-hp engine (Winton model 106) drove the electrical power package. Lacking the nose styling and color of McKeen and GE cars, the flat-faced, Pullman green M300 effectively hid what was inside, namely, a new level of reliability for gas-electric cars that, by the end of production in 1932, had spawned 400 EMC doodlebugs, as they were often called.

But no one can claim EMC's success was just luck, since Hamilton realized that, to get his cars on the road, he would have to make audacious promises on performance, which he was willing to do. At the 1955 Senate hearings, he recalled, "First, we gave them [the buyer] charts and data on acceleration, top speed, and fuel economy. Then we guaranteed our car would meet these conditions. Furthermore, the car was to operate on a scheduled run for 30 days and not be late at the final terminal (over 15 minutes) twice. If it failed this test, we would reclaim the car at no cost to the railroad."

Another of Hamilton's ideas came from his years at White Motors. His plan included service centers scattered throughout the nation so that spare parts could be shipped anywhere within 24 hours. Fortunately, Hamilton's well-conceived plans were in harmony with railroad finances in the mid-1920s. For example, the versatile gas-electrics operated for 50 cents per mile in contrast to $1.25 for a conventional train. Consequently EMC's output soared from 36 cars (in 1925) to 45 (1926) to 54 (1927). But its peak year was 1928 with 105 cars, for a total of 264, or roughly 80 percent of the U.S. market in the company's first five years (1924–28).

The thirtieth car, Great Northern No. 2313, was completed on October 8, 1925. It was 60 feet in length, weighed 32 tons, and had double-ended controls. Power was furnished by a Winton 106A model that delivered 220 hp (in contrast to the 175-hp engine in the CGW car). It is the oldest EMC car still in existence and is now on display at the Mid-Continent Railway History Museum in North Freedom, Wisconsin. It was designated as a Historic Mechanical Engineering Landmark by the American Society of Mechanical Engineers in 2003.

Figure 2.10. Rock Island No. 9012 is an example of an 800-hp, 100-ton gas-electric locomotive that carried two Winton 148 engines in its 51-foot car body. Built at St. Louis Car for EMC in 1929, it was converted to Caterpillar diesels in 1950 and scrapped in 1959. This scene is at Iowa City in June 1940. David Price Collection.

The success of EMC's cars produced economic pressures in two directions: power and fuel efficiency. Most railroads were satisfied at first with a single car train (carrying both people and baggage-express), but as ridership increased and labor agreements changed, a small trailer coach (35 tons) was usually added. By the late 1920s, more or larger trailers (up to 80 tons) were preferred, and consequently power-train requirements rose. Thus engine sizes increased gradually to 400 hp, and in some cases a twin-engine configuration was employed if the gas-electric car was going to pull a short train. The pressure on fuel economy was a result of increasing gasoline demand (and price) due to the proliferation of highway vehicles in the late 1920s. Once again, McKeen and others' engines were modified to burn engine distillate, a heavier fuel than gasoline but much cheaper because it was a limited-use by-product of the crude oil refining method in vogue at the time (Burton cracking process). In simple terms, distillate was halfway between gasoline and kerosene. It was fine for cruising but difficult to use as a starter fuel due to its lower volatility.

With EMC on the radar of the railroad industry, the company continued to grow. In January 1926, Richard Dilworth brought his 16 years of experience with GE's gas-electric program to Cleveland when Hamilton named him chief engineer. Other ex-GE staff came with Dilworth, along with the consulting skills of Hermann Lemp, who had obtained patents for controlling IC engine–generator operation (1914, 1919), but was still not satisfied with the results and continued his developments with support of both GE and EMC. Early electrical engineers had quickly realized that an engine-driven generator was much more difficult to control than the trolley or interurban car, which takes electrical power from an external source without any limit. To control the speed of their traction motors, it is necessary to use only a rheostat (variable resistance device) to decrease the power going into the motor. The rheostat was connected to the motorman's rotary control handle, which generally included six to eight settings (notches).

Chart 2.1. The innovative control system design of Hermann Lemp in 1926 allowed the operator to control both engine and generator with one throttle lever. *Locomotive Operator Instructional Manual.*

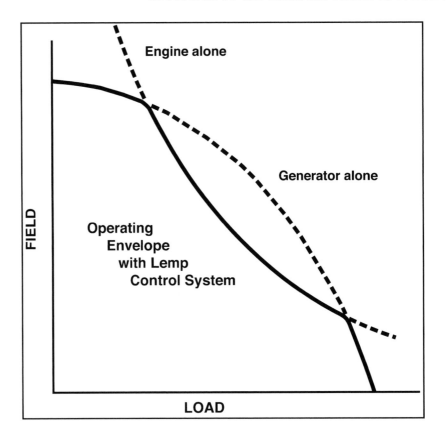

The additional complexity that arises with an on-board generator driven by an IC engine is that it creates what is known as a dynamic system with two degrees of freedom. The latter term means that the motorman can independently control the engine with its throttle and the DC generator with its field strength (surrounding the armature), but that each change in the setting of one component influences the operation of the other. The form of this asymmetry between engine and generator characteristics is shown in chart 2.1 in terms of generator load (amps) and field voltage, overlaid with engine performance expressed in the same physical units. With one curve being concave in shape and the other convex, there are only two balance points where the two distributions intersect. Of course, the generator operates between its field and load limits.

In the beginning, the motorman had to learn by experience (and by listening to the machinery) how to set each control so that the engine drives the generator only enough to supply the power needed by the traction motors. For example, an adjustment with too much throttle would waste fuel, while too much generator load (field) would stall the engine. For a constant-load operation, once the balance condition was achieved, there would be no need for further adjustment. However, this would represent a static system rather than a dynamic one in which the rail car continuously cycles through a start-accelerate-cruise-decelerate-stop sequence, causing the motorman to be perpetually adjusting the controls. Gradually a few simple instruments were provided to give the operator more knowledge of the instantaneous status and thus make his control decisions more efficient. Later the Ward-Leonard control system was developed and would become the standard.

Lemp's overall objective was to make the entire load-balancing process automatic, so that only one control lever would be required to operate the gas-electric car. His third design of 1926 finally actualized the long-sought goal of speed control with a single lever. More-over, it was done with a completely electrical system having no moving parts. It featured split field windings for the main generator as well as separate exciter (generator) for energizing these two fields. By proper choice of the number of turns on the two generator field coils, he could produce a nearly constant power output over a wide range of vehicle speeds. This allowed the operating characteristics of the generator and engine to be virtually matched, so that only one control setting would make both components operate together. This pioneering achievement, also depicted in chart 2.1, illustrates how the engine-generator follows a continuous path that is limited by the generator during one part of the operating range and by the motor in another part.

Lemp's accomplishment represented one of the first important applications of an emerging technical area known as closed-loop control. In simple terms, such a control system depends on a method of sensing a lack of balance between the two competing parameters (throttle and field), and then feeding this information upstream (generator to engine), allowing for an automatic corrective action to eliminate the imbalance. Lemp's patent application summarizes his design this way: "I then provide a secondary field in series with the generator field. Thus the generator load, which at any value of the armature current is proportional to the voltage, will decrease faster than the engine power decreases. Therefore the engine speed will become stable at a value which balances the load." Soon after Lemp's patent was publicized, it became the basis of all controller designs during the era of DC generator use in diesel locomotives, a period that ended in the early 1970s.

During the late 1920s, EMC found itself branching out from exclusively motor car construction. A number of railroads realized that EMC's high-efficiency power package could be retrofitted into other builders' car bodies, especially McKeen's structurally sound cars. At least three lines—Chicago Great Western, Rock Island, and Union Pacific— repowered their McKeens. The 1929 CGW train was actually a precursor to later streamliners. Named *Blue Bird*, the three-car train included a coach and rounded-end observation car (both of McKeen origin), and ran over the Minneapolis-Rochester line that served Mayo Clinic. On another front, Rock Island's Horton, Kansas, shops decided in 1927 to customize a couple of 40-foot steel mail cars and install in each a pair of EMC distillate engines (550 hp). This gas-electric locomotive worked so well that the road converted three more mail cars in 1929 with dual eight-cylinder 400-hp engines. Satisfied with their pulling power (850 tons at 8 mph on a 1 percent grade), the road ordered seven more of these units, three of which were not retired until 1964.

But just as World War I brought a halt to the development of gasoline-powered coaches, the stock market crash of October 1929 plunged the nation into a deep recession and dragged American businesses down with it. In EMC's case, it delivered 92 cars in 1929, but only 58 in 1930. Then the entire market evaporated. But with its enormous talent and experience base, EMC would not fold as did UP's McKeen Car Company and GE's Gas Engine Department. Instead, it would gain a very rich uncle who was interested in the next leap technology—a lightweight diesel power unit to connect with GE's electrical gear.

Figure 3.1. Jersey Central's No. 1000, produced in 1925, was the nation's first commercially successful diesel electric. A product of the General Electric, Ingersoll-Rand, Alco consortium, this 60-ton unit was powered by a 300-hp IR 6-cylinder engine. Its simple, boxy car body typified the first group of switchers used by American railroads to diminish smoke in urban settings. That it was still working at Jersey City in October 1952 is amazing. Louis A. Marre Collection.

CHAPTER 3

THE DIESEL CLIMBS ABOARD

During the three decades when gasoline-powered rail cars were developed, numerous American groups made important advances in diesel engines. Virtually everyone agreed that the compression-ignition concept expounded by Rudolf Diesel in the 1880s would produce a high-efficiency power source. But it became clear after nearly 20 years of effort throughout the world that a reliable and useful diesel engine, especially for locomotive applications, was not going to be built merely by making improvements to the spark engines of the period.

The fundamental technical barrier was a result of the basic nature of the diesel cycle, which operated best at high rpm with many parts subjected to high pressures and temperatures. These requirements in turn necessitated a manufacturing process of much higher accuracy and repeatability than was needed for steam locomotives and even early gasoline engines. Not surprisingly, early diesel designers had used thick-walled structures to contain the pressures and temperatures and operated their machines at modest rotational speeds.

In the 1955 U.S. Senate hearings, Harold Hamilton summed up the situation he saw in the late 1920s like this: "Our gasoline engines weighed about 20 pounds per horsepower and would run at 500 rpm, which is necessary to assure that the generator was small and light enough to fit into our engine rooms. These engines also accelerated well enough to switch cars efficiently in the classification yard. But, after looking around the United States and Europe, we realized that no one had yet gotten a diesel engine down to even 45 pounds, and most were around 60 pounds per horsepower. Now that is fine for large ships, stationary power plants, and heavy mill operations, but it hasn't worked very well in the railroad business." Hamilton had formed this pessimistic opinion by examining the efforts of the pioneering builders of railroad diesels described below.

General Electric

GE's effort during the final years of building gas-electric cars is important because it produced the first diesel designed especially for railroad use. It began when Gas Engine Department head Henri Chatain and electrical engineer Hermann Lemp journeyed to Europe in 1911 to locate a good diesel engine design. They finally decided to buy rights to a design from Germany's Junkers Company. It was an opposed-piston configuration but not like the later Fairbanks-Morse design, which had two crankshafts. The German engine was much more complex, using a single crankshaft with three connecting rods for each cylinder. The center rod connected with the lower piston, while the two other (very long) rods drove a yoke at the top of the cylinder that was connected to the second piston. These extra long connecting rods were the Achilles' heel of the engine because they needed to be stiff and therefore heavy. They were also under high stress loads and prone to break. The primary advantage of the Junkers engine was that it could extract 75 percent more power per cylinder than an ordinary engine. Although the initial GE design represented a significant improvement of the original design, it was still too heavy (105 pounds/hp).

So Chatain and Lemp eliminated the long connecting rods and used parallel cylinders with a common combustion chamber at the top. As in any opposed-piston configuration, each of the two pistons served as the inlet and exhaust valves for the other, allowing for two-cycle operation. Moreover, the large cowl covering the two-cylinder combustion chambers made the engine appear to be a V-4. With an 8-inch bore and 10-inch stroke, the GM-50 eight-cylinder machine could deliver up to 250 hp at 550 rpm. The test vehicle for this engine, a truncated motor car body (rounded nose) in which the engine sat over its power truck, made its initial run in late 1916 (or early 1917). The car body would later be resurrected for an Ingersoll-Rand engine test.

The first new unit carrying a modified engine (GM-50A2) was mounted in a steeple-cab car body and sold to Brooklyn's Jay Street Connecting Railroad. A second such unit was delivered three months later to Baltimore, while the final unit of this series was an armor-plated vehicle for the U.S. army that, by all accounts, was never operated except in GE tests. Sadly, this experimental program to construct a railroad-worthy diesel engine was a complete and colossal failure, with most of the blame placed on the failure of high-pressure seals around the injector valves, which at the beginning had a life expectancy of only eight hours. This major debacle brought considerable internal disfavor to the Gas Engine Department, and it was soon disbanded. Most of the technical staff moved on, but Lemp and A. F. Batchelder remained to continue the program.

Ingersoll-Rand

An often-overlooked pioneering builder of diesel engines was Ingersoll-Rand, which also teamed with GE for electrical gear. As its hyphenated name suggests, the IR Corporation resulted from a merger of two companies in businesses related to the mining and construction industries. The Ingersoll Rock Drill Company was founded in 1871 by Simon Ingersoll to market a steam-powered drill he had invented. He merged his company with Sergeant Drill in 1887 and moved from New York City to Phillipsburg, New Jersey, in 1903. Only two years later, Ingersoll merged again, this time with his largest competitor, Rand Drill Company. The latter had begun work on reciprocating machines in 1898 when it began building air compressors at Painted Post, New York, under the name Imperial Engine Company. This work involved a line of spark-ignition engines for natural gas pipeline pumping. These machines were able to run in remote areas, since they could tap fuel from the pipeline. Some of these innovative machines featured a common crankshaft onto which both engine and compressor pistons were connected.

Ingersoll-Rand became interested in switching to diesel power in the early 1930s because these engines could burn heavy oil, which was much cheaper than gasoline or natural gas. Early experiments between 1914 and 1916 were discontinued due to World War I but were revived when William T. Price joined IR in 1918. Price had not only started his own engine company but also had experience with other engine builders. Indeed, he had patented a new combustion chamber in 1917. His device, sitting atop the cylinder, promoted more rapid burning by forming a swirling flow that thoroughly mixed the fuel spray and air. In late 1918, IR introduced a line of single-cylinder horizontal stationary engines, denoted as the PO series (for Price Oil). At this time a careful delineation was drawn between an *oil* engine and a *diesel*. The former injected both high-pressure air and fuel droplets whereas the latter (also called solid injection) involved the injection of a fine mist of fuel directly into the top of the cylinder when the piston was near TDC (maximum pressure and temperature, chart 1.2).

The PO engine was delivered in sizes ranging from 55 to 150 hp. The largest weighed 24 tons and lumbered along at 200 rpm. But it was also reliable, as illustrated by one company that kept its PO running continuously for eight months and 20 days! These stationary units showed clearly the diesel's potential, but their bulky 320 pounds per horsepower was certainly of no interest to builders of vehicles. After Price's sudden death, one of his primary lieutenants, George Rathbun, became chief engineer of IR's Oil Engine Department. He went on to develop a line of multi-cylinder inline diesels known as the PR series (Price-Rathbun). Starting in 1920, they were produced in sizes from 50 to 1,200 hp and became popular for marine quick-reversing power and general stationary use.

Ingersoll-Rand's entry into railroad diesels came about as a result of serendipitous circumstances at General Electric. With its Gas Engine Department gone, GE management told Batchelder and Lemp that, if they wanted to continue building railcar electrical gear (including improved control systems), they should buy an engine from the outside. When they could find none, Lemp wrote detailed specifications for a new engine and circulated copies to all American diesel builders in 1920. He wanted a 300-hp machine with high reliability, limited size and weight, and low vibration. The response to his solicitation—challenge, if you will—was a resounding silence, with most companies quick to dismiss the engine as impossible to build. The only response came from the brash bunch at IR, who believed that their PR engine could be modified to achieve the specs. By 1923 they had produced a simple six-cylinder inline engine with 10 × 12 cylinders, an 11:1 compression ratio, Price Combustion Chamber, and droplet injection. The crankcase was a heavy box to which the cylinders were bolted. They were jointed at the top with a heavy manifold that gave exceptional rigidity to the engine. Weighing 65 pounds per horsepower (within GE's specs), the engine's structure was designed conservatively (for low stress levels). Having a nominal maximum output of 300 hp at 550 rpm, it could go up to 325 hp at 650 rpm for short periods.

Lemp and Batchelder were satisfied with IR's efforts and immediately began sifting through GE's storage yard for a used car body to recycle. They finally chose a double-ended boxcab used in the 1917 tests of GE's ill-fated diesel that brought down the Gas Engine Department. Its final appearance included the original rounded nose, single-ended controls, and large radiators on the roof. The historical unit, America's first commercially feasible diesel-electric locomotive, was numbered 8835 (its serial number) and lettered simply Ingersoll-Rand Co. It was formally rolled out on February 28, 1924, to an audience that included representatives from B&M, B&O, LV, NYC, Reading, and Pennsy, plus technical reps from IR, GE, and Alco, which would construct car bodies for production models. All of the railroad representatives were impressed but naturally skeptical. However, they welcomed IR's decision to loan the unit to any line that wanted to try it.

The diminutive boxcab spent the next 13 months, starting in June 1924, strutting its stuff on ten railroads and three industrial lines. For its appearance on the NYC's West Side line in Manhattan, it was decorated with several GE emblems and the names of the three participating companies. Later the 60-ton engine showed exceptional strength when, after much groaning and straining, it moved a 93-car train on level track. It also scored more points when, halfway through the demonstration period, it was returned to IR for inspection and repair. Astonished engineers found that, after seven months of work, there was only minimal wear on pistons and crankshaft bearings. The engine was closed up without a single new part and sent out again. However, when No. 8835 completed its demonstration tour, it returned to IR as a plant switcher in July 1925 and was retired in 1926, its car body returned to GE's storage yard and its engine to the IR lab as a test bed.

The sales effort on the new locomotive fell to IR due to a strange twist of corporate

Figure 3.2. In April 1930, Rock Island received a GE-IR switcher tailored for smoke-free service around Chicago's LaSalle Street and Union stations. Numbered 10000, the unit featured a small Ingersoll-Rand 300-hp diesel running a DC generator to charge a bank of 240 batteries that supplied traction motors. Called a "bi-power" machine, its modern descendants are known as hybrid units (see chapter 12). This pioneering unit would not be retired until 1950. C. W. Witbeck Collection, courtesy of David Price.

politics. At this time GE management was fixated on electric locomotives and components while Alco was still going strong as one of the three major steam locomotive builders. Constructing diesel car bodies would continue as a quiet sideline business at Schenectady for years. Thus the marketing campaign fell to the IR sales staff, led by general manager L. G. Coleman. They began by ordering three stock units (300 hp) in July 1924. Alco, GE, and IR ironed out the final design details that saw little external change except that the unit lost the round-nose and gained a separated operator's cab, which was partitioned from the engine room to reduce noise and heat. Alco shipped the three car bodies to GE's Erie plant in late 1924, while the IR engines arrived soon thereafter, and the final assembly of America's first production unit was completed on June 1, 1925, and put through a rigorous testing program. The other 300-hp units plus a twin-engine 600-hp version were finished in a more leisurely fashion because there was still one component missing—a buyer.

What happened next was completely unanticipated. While major trunk lines, led by steam era executives and mechanical staffs, generally scoffed at these small machines, there arose an unexpected market due to the dominance of the steam locomotive. In the late 1920s, noise and air pollution were beginning to strangle large urban areas, and these quiet and smoke-free yet powerful machines had appeared at just the right moment in time. Of course, Manhattan was a primary target for cleanup after the borough banned steam locomotives in 1923 with the Kaufman Act, with a similar ban soon in place for Chicago's Loop District. The Jersey Central quickly installed a 60-ton IR diesel (No. 1000) on its Bronx car float in October 1925, followed by two identical units acquired by B&O and Lehigh Valley for car float operations on Manhattan's West Side. A 600-hp Long Island unit scored publicity points in December 1925 by running all the way from the Erie plant to New York City pulling a seven-car train. The 537-mile trip took nearly 29 hours, although the switcher was fitted with temporary gearing for higher speeds. The diesel used only 473 gallons of fuel and 5 gallons of lube oil. Total cost: $26.15! In another direct comparison with steam, the CNJ unit cost $301 to operate during its first six months against $1420 for the last six months of steam operation.

With demand exceeding supply in late 1925, IR quickly began another stock program that continued for a year. Eight units were ordered in December, six in February, three in

Figure 3.3. The nation's first C-C units appeared in 1936 when two 1,800-hp, modified boxcab designs were constructed by the GE-IR-Alco group for transfer service on the Illinois Central in Chicago. J. Parker Lamb Collection.

April, and six in October. Thereafter, units were built as ordered except for another stock order in October 1930. The bulk of IR's total production of 116 units was built between May 1925 and October 1937. There were 106 units using the standard 300-hp engine either singly (80) or dually (26). The other ten units used engines in the range of 400–900 hp. Car body styles included 94 boxcabs, 15 single-engine hood units, and 7 twin-engine hoods. Among the unusual variants from the common boxcab B-B configuration were 45 NYC bi-power and tri-power units (fitted for battery and/or electric operation), a 2-D-2 boxcab (750 hp) for NYC, and a pair of 1,800-hp twin-engine C-C hood units built for Illinois Central in 1936 (first use of these trucks), and finally a low-nose end-cab switcher for a steel mill in India. Significantly, sales to industrial customers represented one-fourth of IR's total output. Due to the nature of their operation, most of these units outlasted their brothers bought by common carriers.

Ingersoll-Rand's conservative design philosophy led to a fleet of reliable, long-lasting units, but by 1934 its engine designs were being overtaken by those of Cummins and Cooper-Bessemer, forcing a number of IR improvements in 1935, including higher horsepower. However, only eight units were produced with the newer engine, the last being a pair of switchers for Milwaukee Coke and Gas Company in December 1936 and October 1927. Thus IR's 12-year run as a diesel-electric locomotive builder came to an end, but it still represented the nation's first step toward railroad dieselization.

Westinghouse Electric

As noted earlier, during the last two decades of the nineteenth century, Westinghouse Electric & Manufacturing Company (WEMCo) rose quickly to challenge the older General Electric Company. Together the two became America's leading developers and build-ers of railroad electrical equipment. Much of that was dedicated to mainline electrifica-tion, primarily in the eastern third of the nation. WEMCo, a latecomer to the diesel party, was escorted to the proceedings by an unlikely suitor, William Beardmore, a well-known company from Glasgow, Scotland. Led by Alan E. L. Chorlton, Beardmore was able to design and construct an extremely light diesel engine (16 pounds per horsepower) for use in a lighter-than-air dirigible during World War I. The weight reduction was achieved by using aluminum castings with thin walls. Unfortunately, the crash of one of the airships led to some criticism of Beardmore's work, although the Ministry of Aviation never officially blamed the company.

In 1922, Beardmore constructed some diesel-powered railcars for the London Midland and Scottish Railway using a heavier derivative of the airship engine that weighed 22 pounds per horsepower. Soon the Canadian National Railways became interested in Beardmore's design and ordered engines for nine passenger motor cars that the road planned to assemble in its own shops after ordering all components from outside. The Thomson-Houston company in Great Britain was chosen to build the electrical gear for seven 60-foot cars that used a four-cylinder 185-hp engine, while Westinghouse supplied corresponding apparatus for an articulated passenger car body that was powered by two eight-cylinder engines, rated at 600 hp and 600 rpm. Curiously, this design was based on the use of battery power, but in actual usage the system failed badly and the cars were converted to traditional diesel-electric drive using Westinghouse gear. After subsequent failure of the control system, WEMCo also installed its new torque control system, since Lemp's early patents were still held by GE.

These locomotive-related activities caused Westinghouse engineers to become inter-ested in developing their own engine design based on the Beardmore machines. After signing an agreement for patent licenses in mid-1926, the WEMCo team began upgrading the Scottish design. In the meantime, Beardmore production engines (six-cylinder, 300 hp, 800 rpm) were imported and sold in the United States with a builder's plate that read "Westinghouse-Beardmore, built at South Philadelphia Works." As long as the engines worked well, no one seemed to be bothered by the mythical nameplate. To order car bodies for its locomotives, Westinghouse had only to call Baldwin's nearby Eddystone, Pennsylvania, plant. In September 1927, a pair of car bodies resembling the forward-third of gas-electric motor cars were delivered by Baldwin to WEMCo's East Pittsburgh plant. When completed in September 1928, the initial Baldwin-Westinghouse locomotive was composed of two four-wheel units coupled in a B+B arrangement, each carrying a 300-hp engine. After testing, the units were delivered to the Long Island railroad in May 1928, soon earning the nicknames "Ike" and "Mike."

The next project represented a leap from these small units to a pair of very large machines. It was a quintessential case of construction by committee that arose this way. In 1928 Canadian National contracted with Canadian Locomotive Company (CLC, Kings-ton, Ontario) to assemble two car bodies whose components were fabricated by Common-wealth Steel using detailed designs by Baldwin. Powering the large 2-D-1 running gear on each unit was a 1,300-hp V-12 Beardmore engine (800 rpm), while electrical gear was built by Canadian Westinghouse (Hamilton, Ontario) using designs by WEMCo. Together the articulated locomotive, No. 9000, weighed a whopping 335 tons, and the railroad's PR staff was quick to provide widespread coverage of what was then the world's largest diesel electric. The machine proved to be a reliable locomotive with performance comparable to a CN 4-8-2.

While the 9000 was taking shape at Kingston, two smaller units were on the erecting floor at the East Pittsburgh plant. Both were 55-ton B-B boxcabs powered by the standard 300-hp Scottish engine and supported on Baldwin trucks. Despite the similarity of their appearance, their car bodies were fabricated in different plants, one at Baldwin and one at WEMCo. This exercise was an attempt to gauge the comparative cost of internal vs.

Figure 3.4. Westinghouse Electric & Manufacturing Company followed the logical path and started small with its diesel locomotive development in 1928. This diminutive 600-hp combo for Long Island switching service was quickly dubbed "Ike" and "Mike." Westinghouse Electric, J. Parker Lamb Collection.

vendor construction. Not surprisingly, Baldwin was cheaper. After completion, the two units ran demonstrations in various industrial settings, with the Baldwin model later sold to a Western Electric plant, while the Westinghouse-built unit was assigned as a plant switcher. Meanwhile, the first production models (Mike and Ike) required new engines after a year of service with 18 percent availability. Most observers chalked this up to teething problems with the Beardmores. In 1929 WEMCo constructed a specially protected unit for an Armco Steel mill in Butler, Pennsylvania, 35 miles from Pittsburgh. This engine carried a heavy underframe and, for protection against high heat in the mill, received special insulation and high-temp glass windows as well as dual controls.

Westinghouse engines also powered 19 passenger motor cars in 1929 and 1930 that went to Pennsy, Reading, Erie, and CN. Car bodies were constructed by a variety of builders including Pullman, Bethlehem Steel, St. Louis Car, and CLC (Ontario). The CN cars were the last to use imported Beardmore engines, as WEMCo's new prime mover was finally ready to install. The new diesel, built under the direction of David Morgan, incorporated a number of physical and operating changes, including a larger bore (8.5 to 9 inches), increased speed from 800 to 900 rpm, and completely redesigned auxiliaries (injectors, governors, manifolds). The new engine provided an extra 100 hp over the standard Beardmore 300.

WEMCo engineers, led by David Hershberger, also tackled the poor visibility of the boxcab car body design by decreasing the width of the body near the top, thus giving the cab crew a clear view to the front (much like later hood units). The first road to purchase this new design was CN, which opted for a 70-ton, 400-hp unit in 1930, assembled by CLC. Soon plans were made to offer a standardized line of locomotives using the new 9 × 12 engine and the higher visibility cab. Three demonstrator units were ordered: two 70-ton, 400-hp end-cabs, and a center-cab 800-hp (dual engine) configuration weighing 110 tons. However, before any of the demo units were completed, WEMCo sold five 70-tonners to various steel mills in Pennsylvania, Ohio, and Michigan. After the 800-hp center-cab unit (No. 23) ran many miles on demonstration tours in 1930–32, the Northampton & Bath purchased it in 1933.

Although sales dropped to zero at the height of the Depression, Westinghouse continued to refine its designs in anticipation of a market rebound that never came. In 1932, it

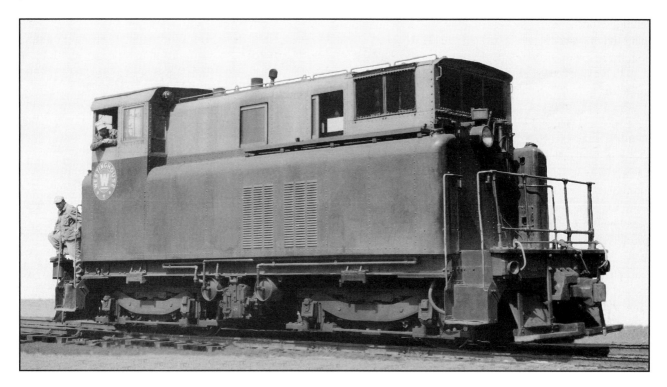

Figure 3.5. In 1929, Westinghouse introduced a line of switchers using its newly designed diesel engine. Seventy-ton, 400-hp No. 8 was in the first batch produced, and it carried the company's new design for a "visibility cab." By the time it was photographed in March 1962, it had been reengined with a Cooper-Bessemer diesel and was the East Pittsburgh plant switcher. Jerry Pinkepank, Louis A. Marre Collection.

began assembling Pennsy P-5 electric power using Baldwin-built car bodies and turned out an occasional switcher. One of the highlights of an otherwise drab period for WEMCo locomotives was the April 1935 appearance of New Haven's *Comet*, a three-car, shovelnose streamliner with power cars at each end (using 400-hp Westinghouse engines). The car bodies were constructed by Goodyear's Zeppelin Division at Akron, Ohio. With sales continuing to decrease, WEMCo consolidated its locomotive operations in a smaller plant and waited for the end, which came with the delivery of three end-cab switchers to Great Lakes Steel in January 1937. Thereafter, Westinghouse became a major supplier for other builders, such as Alco, Baldwin, and even EMC.

General Motors

Almost three decades had passed since William McKeen and Edward H. Harriman had set into motion the development of American's first gasoline-powered coaches. As an economic and morale booster for a Depression-weary nation, the 1933 Century of Progress Exposition was organized and held in Chicago. Diesel-powered rail vehicles had been operating since 1917 (GE) but had not seen any commercial success until Ingersoll-Rand's emergence as a builder in 1924. Yet the newest type of IC engine was still in the infancy of its technical development, and railroads were still devoted to dependable steam locomotive technology.

Not surprisingly, one of the major exhibitors at Chicago in 1933 was General Motors, which decided to provide diesel-electric power for its exhibit of a Chevrolet assembly line. Sitting on a gleaming checkerboard tile floor behind brass rails within a completely glassed-in corner room, two enormous and spotless engines droned on day after day with relatively little noise and no exhaust (it was piped away). Then a strange thing happened. The curious crowds around these sturdy-looking machines began to increase, and soon the power corner became a major attraction in its own right, as people marveled that such immense power could be generated in a small and relatively clean room. Moreover, the majority of visitors who observed the diesel-powered generators were probably surprised that the name cast into the crankcase covers was not General Motors but Winton.

The man responsible for putting these two engines in the Chicago exhibit was Charles Kettering, one of America's most honored technical leaders. He was a man of enormous intellect, intuition, and enthusiasm for exploring the unknown, especially for solving technical problems. Both Charles and his son Eugene would help GM become the dominant player during the first 50 years of American dieselization. The elder Kettering was born in Ohio in 1876. His poor eyesight and his need to be self-supporting during college delayed his receipt of an electrical engineering degree from Ohio State University until he was 28. His first job after graduation was with National Cash Register in Dayton. Although lasting only five years, his tenure at NCR was marked by many innovations in developing new electrical devices. Not content with working for a large company, he founded his own R&D firm in 1909, known as Dayton Engineering Laboratories Company. Under his hands-on leadership, Delco (as it became known) soon developed a completely new type of automobile starting, ignition, and lighting system. Considered a major advancement emblematic of luxury cars, it was installed in the 1912 Cadillac, and Boss Ket, as his employees fondly called him, began a lifelong relationship with GM. In 1916 Kettering sold Delco to United Motors, a GM subsidiary. Moving to Detroit, he assumed the role of the automaker's premier tinkerer and visionary.

As the diesel engine evolved in the early 1910s, GM's technical staff had maintained close contacts with German and Swiss companies that were doing such developmental work. However, nothing in this early era caused GM people to give this concept much attention. Later, at a 1927 Midwest meeting of the Society of Automotive Engineers, there were formal presentations about the future of diesel-powered trucks and locomotives. After hearing various viewpoints, most GM technical people were somewhat pessimistic about the diesel's future—except for Charles Short, who had once worked with the experimental diesel program of General Electric's Gas Engine Department.

Soon Boss Ket decided it was time for a bit of personal investigation, an enduring characteristic of his hands-on approach. In early 1928 he ordered a yacht from the Defoe Shipbuilding Company in Bay City, Michigan. It was 105 feet long and powered by a 175-hp, four-cycle Winton diesel. While others enjoyed the deck, Kettering spent many hours below, studying the engine's operating characteristics. Generally satisfied with the engine's performance, he soon learned that the operation of the fuel injectors was inconsistent and that the four-cycle design produced an extensive amount of heat. Despite this, he was convinced that the diesel engine had a bright future in American transportation, but only if these shortcomings were fixed.

At this point Kettering went to GM's executives and argued strenuously in favor of acquiring an American diesel engine company. Given the opportunity to make a formal proposal, he visited a number of companies, including Trieber, Cummins Engine, and Winton Engine. He decided on Winton, and the acquisition was made in 1929. Soon he was doing more seaborne research after GM ordered a number of larger yachts for some of its executives. The new boats were 65 feet longer than the original one and were powered by two 600-hp Winton diesels. As usual, Kettering did some customizing to this vessel, including the design for an electromechanical stabilizer (later marketed by Sperry Company), as well as an engine synchronization system that allowed both propulsion screws to turn at the same rpm, producing much less discomfort for passengers. After some shakedown runs, Boss Ket organized a group cruise to South America (Galapagos Islands). Included on the passenger list was his son, who wanted a vacation from college but also an opportunity to learn more about the big engines under his father's tutelage.

Needless to say, both Ketterings closely observed the operation of a new unit-injector. Designed by Winton's Carl Salisbury, it used separate pumps for each cylinder, thus dispensing with long lines of high-pressure piping strung around the engine. Unfortunately, the injector's performance turned into more trouble than was expected. CFK noted in a letter to a colleague at GM, "I do not think we ran at any time over 24 hours without breaking an injector nozzle." Ancillary troubles included rough running (promoting discomforting vibrations) and stalling out (when idling). More seriously, engine valves were burned from excessive heat, and this required repair stops for valve grinding. One day, during a storm, the Ketterings worked feverishly below deck to keep the propellers spinning. Finally during one 30-hour stretch of babysitting, their close attention and mechanical insight produced the nugget of gold for which every diesel builder had been searching. They discovered that the root of the problem was the kind of material used for the cylindrical injector body and the plunger pump inside.

It was common to use stainless steel for applications in such hostile environments. The metal was durable and could be drilled to form the tiny nozzles for droplet injection into the combustion chamber. But stainless steel was a very poor bearing material due to spalling (minute gouges on its surface). This prevented the plunger from moving uniformly and thus destroyed the timing of the droplet spray, making the engine run roughly. Once this problem was clearly identified, the solution was deceptively simple: use different materials for the barrel and plunger, and move the injectors to a cooler part of the engine. These fixes, made after the Ketterings were back at the Winton labs, gave diesels the smooth running that everyone had expected. After this exhilarating experience, Gene decided that college could wait. He wanted to get to work with Salisbury at Winton and develop a new and improved engine.

General Motors was now firmly behind diesel engine development, having just received a U.S. navy contract for a diesel-electric generator set. Soon work was begun on two fronts. In Detroit, Boss Ket and F. G. Shoemacher set up a lab to optimize two-cycle operation, using a relatively simple single-cylinder engine, the simplest device that contained all necessary elements. In Cleveland, Gene Kettering and Salisbury built another one-cylinder machine to develop improved pistons and injectors. During this effort, Gene built and tested 48 variations of piston shapes before finding the one he thought was best. Later the best features of both test engines were combined into a super prototype that was able to produce 75 hp (or 20 pounds per hp). The final navy engine also included the ultimate in interchangeability. Each piston-cylinder power pack could be removed from the main engine block for easy replacement. With this successful design, Winton people could now envision a large production engine on the horizon.

But Boss Ket, the archetypal risk taker, suddenly stepped in and decreed, "Let's build an eight-cylinder version and use it at the Chicago exhibition!" Most company people shook their heads in disbelief. How could he expect to go from this small prototype to a full-size engine that quickly? But, of course, with such dedicated engineers as Gene and his Winton counterparts, an around-the-clock companywide effort allowed them to build and partially test the two Winton Model 201 engines that went to the Chicago exhibition in 1933. Both Ketterings later recounted some of the harried times during the building of the 201. CFK said in a 1951 conversation, "When we first started up the new engine, oil, water, and fuel spewed forth from almost every place except where it should." Of course, this required the installation of new seals and gaskets, a typical process with high-pressure machines. On the other hand, Gene, who was delegated to lead the technical team that babysat the machines in Chicago, remembers that they all slept in a room just below the engines, in case there were unexpected problems. He recalled, "The shop boys often worked all night to get the machines ready for the next day. To mention all the parts we had to replace would take too much time, but I do remember that there was no trouble with the oil pan dipstick!"

It is clear from hindsight that GM's success, in an area where many others had failed, was due to its vast financial resources as well as its dedicated staff of diverse and highly

Figure 3.6. The engine block of the 600-hp, two-cycle Winton 201A inside the restored *Flying Yankee* car body illustrates the compact configuration of the nation's first diesel for railroad use, brought to commercialization by the pioneering work of Charles Kettering and his son, Gene. Photo at Claremont & Concord Railroad shops in June 2003. Jim Shaughnessy.

competent technical people. Both of these critical components were focused completely on the remaining roadblocks to full commercialization of the compression engine some three decades after Rudolf Diesel nearly blew himself up while testing an early model.

Most historians have given Charles Kettering the lion's share of credit for getting GM into the locomotive business, and this well-known anecdote from one of his interviews suggests how brashly he wielded his influence toward this goal. When GM president Alfred P. Sloan became aware that EMC was receiving an increasing share of development funds, Sloan said in a phone call, "Ket, we've got to throw this thing out." Instead of a direct reply, the undaunted engineer applied a combination of psychology and logic. He immediately requested an additional $500,000 for more research on the locomotive diesel. Sloan replied, "You know damn well you can't build a locomotive for that kind of money." Quickly Kettering said, "Yes, but I think if you get that much money in it, you'll be likely to finish it."

Another GM Acquisition

The 1955 recollections of Harold Hamilton provide details of the transactions that put General Motors into the driver's seat during the first decades of American railroad dieselization.

After Winton began making our gasoline engines in 1923, George Codrington (Winton general manager) and I had many discussions about the future of diesel engines, including what operating problems they had encountered with their marine diesels. It was clear that they knew what needed to be done, but there was really no strong impetus to move on it, plus it would be quite costly. However, around 1928 Winton built some diesels for yachts ordered by GM executives. Soon Codrington mentioned that GM was showing a high level of interest in acquiring Winton. He and I agreed that such a move would be great for EMC, since it would give Winton the necessary capital and technical support to make some major improvements in their diesel engines. During the Winton-GM negotiations I became acquainted with Mr. Kettering and found that he and I had much in common. His enthusiasm and persuasiveness were quite exhilarating. In fact, he was one of the most interesting engineers I ever met.

In the fall of 1930 Codrington, who was now president of Winton, mentioned that the GM people had inquired about buying EMC. He asked me to think about it and mentioned that he would talk to me later. Since we [EMC] were a privately held corporation, there were only a few people to contact, and they all agreed that this would be a positive move for us. The next time Codrington came to see me, he soon got around to asking, "How much do you want?" Events moved quickly from there, and everything was settled in December. Our little shoestring company was now a subsidiary of General Motors. The future looked bright.

Hamilton also remembered a discussion with a GM executive prior to the acquisition.

He said, "Now you [EMC] don't own anything; you have a quarter million dollars in spare parts and a bunch of cash. Just what are we doing buying this outfit?" That was sort of unexpected, so I took a deep breath before replying, "You are buying some know-how and the potential base from which we can establish a new industry." Fortunately, he was satisfied and that subject didn't come up again.

CHAPTER 4

STREAMLINED TRAINS

While Winton and GM were struggling to finish the first Model 201 diesel engine, there were some high-level visitors at the Electro-Motive offices in Cleveland. Three Union Pacific executives, chairman Averell Harriman (son of Edward), president Carl Gray, and vice president E. E. Adams, sat down with Hamilton and Dilworth to talk about a new idea. As Hamilton recounted in the 1955 Senate hearings,

> They [UP] had been pioneers with their McKeen cars and had later bought many units from us. So they knew a lot about electrically driven rail cars. They had determined that the next step for them was a high-speed train, not very long, that could be a competitor for buses and the new airplanes that had begun operating in the West. It had to be comfortable and, most importantly, be able to run at 120 mph (about the speed of contemporary planes).
>
> After many design sessions with the car builder Pullman-Standard, the railroad, and our people, we finally cooked up a probable design for the external appearance. An airplane designer then made a wooden model and took it to the wind tunnel at the University of Michigan for testing, since we had no idea of the power we would need to run this train. When the design was finished, the railroad announced publicly that they were going to build it, so we [EMC] entered an entirely new type of activity.

What had been cooked up, as Hamilton put it, was a three-car set made of aluminum and powered by a 600-hp V-12 Winton Model 191A, a distillate engine running at 1,200 rpm. UP's Adams had selected the older engine over a brand-new version of the Winton 201 being designed. Obviously, he was more comfortable betting on the proven horse than going with the young long shot.

In virtual lockstep with Union Pacific's planning was that of Ralph Budd, president of the Burlington Route, and his team. They had taken a close look at the 1932 development of an unusual motor car that ran on rails but used rubber tires. Designed by the French firm Michelin, it was constructed by the Edward G. Budd Company using an entirely new method of fastening sheets of stainless steel, known as shot welding. Burlington also wanted a short, fast train, but soon decided that theirs was to be constructed of this shiny new material rather than aluminum.

During the Chicago exhibition, Ralph Budd had discussed at some length the progress on Winton's new diesel with Hamilton and others. Hamilton, the consummate salesman,

Figure 4.1. Icons of
American industry gather
in Philadelphia before
formal christening of the
Burlington Railroad's
Zephyr in April 1934.
Corporation presidents
include (from left) William
Irvin, U.S. Steel; Alfred P.
Sloan, General Motors;
Gerard Swope, General
Electric; Edward G. Budd,
Budd Car; and Ralph
Budd, CB&Q. Marguerite
Cotsworth will soon swing
the ceremonial champagne
bottle. Mark Reutter
Collection.

assured him that the new engine would last about 80,000 miles before the power packs required changing. He concluded, "I believe we can get an engine built for you soon enough not to delay the completion of your train." The brash Budd replied, "OK, I'll take it," and signed an order for a pair of eight-cylinder, 600-hp Winton Model 201As on June 17, 1933, only a month after UP's order for the distillate engine. When Charles Kettering characterized Budd as "a very nervy railroad president" for his decision to go with an uncompleted machine, Budd replied, "I know that GM will stick with this until it works. I'm not taking a chance at all."

CB&Q's design process also involved wind tunnel tests at Massachusetts Institute of Technology showing that, above 40 mph, there was an increasing level of aerodynamic drag that must be overcome by the engines. But the power plant wasn't the only innovation included in the train. Its car body contained the same interlaced tubular structure that Edward G. Budd had built for McKeen motor cars 30 years earlier as a young superintendent for Hale & Kilburn. However, an additional strengthening mechanism was the addition of fluted (corrugated) side panels. The two new trains hit the road in early 1934, with UP's M-10000, later named the *City of Salina*, rolling out of Pullman-Standard in February, and Burlington's No. 9900, named the *Zephyr* (after the Greek god of wind), pulling away from Budd's plant in April for testing. Despite the fact that two design groups, working 800 miles apart, were responsible for their different external appearances, major measurements were nearly identical. With two cars trailing the power unit, each train was approximately 200 feet in length and weighed about 90 tons (equivalent to a heavy-weight Pullman). There were obvious differences in the appearance of the power units, with the M-10000 having a large nose-shaped radiator below its high-mounted cab compartment, whereas the shovelnose 9900 carried its radiator at the top of the car, above the cab windows.

With their small weights and plenty of horsepower, both motor trains could almost fly. In early tests, the M-10000 raced up to 111 mph—approximate cruising speed of a Ford TriMotor airplane—and made a deep impression on the UP by topping Sherman Hill without a helper. A few months later, No. 9900 hit 100 mph on its first test run out of

Figure 4.2. CB&Q's pioneering *Zephyr* speeds past the tower at Berwin, Illinois, in September 1936. David Price Collection.

Philadelphia, and later it averaged 80 mph as it zipped across Ohio and Indiana on Pennsy rails to Chicago. During demonstration runs, the new trains constantly impressed on-lookers, whether they stood beside them between runs or stared in open-mouthed astonish-ment as they raced past. But it was the visionary Ralph Budd who provided the fanfare that announced a new era in train travel and, implicitly, a new era in locomotive propulsion.

On May 26, 1934, the month-old *Zephyr* slipped out of Denver Union Station in the predawn darkness (4 AM) and headed east. As soon as it was away from the city, it owned the CB&Q main line to Chicago. With track turnouts spiked and crossings guarded, Budd's railroad shepherded the silver bullet on a historic nonstop trek. At a few minutes past 6 PM, the short train raced past Halsted Street Station in Chicago, having traversed 1,015 miles in a little over 13 hours at an average speed of nearly 78 mph. It soon eased over temporary rails onto the Chicago Fairground's stage in full view of a roaring and ecstatic crowd.

This triumph by American technology was a watershed event in the evolution of the diesel-electric locomotive. In his inimitable style, *Trains* magazine editor David Morgan would characterize the significance of this event to Harold Hamilton and his company many years later when he wrote, "Electro-Motive had pulled on long pants."

With this newfound national visibility, EMC received not only widespread acclaim but also orders for more streamlined trains and the diesel locomotives to pull them. In Febru-ary 1935, a near duplicate of the *Zephyr* was created for a joint Boston & Maine/Maine Central train named the *Flying Yankee*. Both it and the pioneering CB&Q train were preserved after retirement, with the former being rebuilt into operating condition by the Flying Yankee Restoration Group of Claremont, New Hampshire, while the *Zephyr* is displayed at Chicago's Museum of Science and Industry. In April 1935 the Burlington received two more power cars (Nos. 9901 and 9902) to inaugurate the *Twin* (Cities) *Zephyrs*, and in October No. 9903 was delivered for use on the *Mark Twain Zephyr*.

Illinois Central became the fourth railroad to get an EMC-powered, articulated stream-liner with its *Green Diamond* in March 1936. Strongly resembling the UP's M-10000, the Pullman-built five-car emerald-hued articulated was expected to be powered by an

Figure 4.3. The New England version of Burlington's *Zephyr* was its twin, the *Flying Yankee*, a joint Boston & Maine–Maine Central train between Boston and Bangor. In late 1944 it became B&M's *Cheshire* and is shown (top) at Track 22 in Boston's North Station before its run to White River Junction, Vermont, in June 1952, five years before its retirement. Lower photo (June 2003) of the *Yankee* restoration in New Hampshire permitted a look at the interior skeleton of the historic stainless-steel streamliner. Jim Shaughnessy.

Figure 4.4. Led by a 1,200-hp unit with a B-2 wheel arrangement, Illinois Central's four-car articulated train, the *Green Diamond*, began Chicago–St. Louis service in April 1936. Eventually it became too small for prevailing traffic levels, and in 1947 it was renamed *Miss Lou* and assigned to the New Orleans–Jackson, Mississippi, route. Shown at Brookhaven, Mississippi, in August 1948. C. W. Witbeck, Louis Saillard Collection.

eight-cylinder Winton, but after initial tests showed this to be inadequate, it was replaced by a new model, a V-16 version of the 201A producing 1,200 hp. A month after IC's train hit the rails, the company produced two motor cars (only baggage and RPO) for Seaboard Airline branch lines. These shovelnose cars (Model AA) were built by St. Louis Car and powered by the eight-cylinder 201A. They represented the first use of diesels in an EMC motor car.

Also in 1936, Union Pacific began expanding its fleet of Pullman-Standard trains and ordered new power cars M-10001 through M-10007. All lead units, powered by the V-16 1,200-hp engine, were coupled permanently to a booster unit in recognition of increased train lengths on the extremely popular streamliners. The first booster carried only a 900-hp engine, but the remaining ones also used the larger engine, so total power was now at 2,400 hp. Within three years, almost all of these cars were modified to produce two sets of A-B-B power (3,600 hp) for the *City of Denver* trains. The lead units received a longer nose (with frontal facing radiator) and a new turreted cab, while the number pattern for these multiple-unit sets was CD-06A, -06B, and -06C and a similar CD-07 series.

This somewhat unorthodox numbering scheme was the harbinger of a protracted disagreement during the 1930s between labor unions and management over the question "What constitutes a locomotive?" With railroads scrambling to embrace diesel power, whose electrical drives allowed a number of separate power units to be controlled by one engineer, labor leaders became concerned about the potential loss of jobs, especially those of firemen. One union proposal would have required a fireman in each locomotive unit to perform en route maintenance. Moreover, diesel operations showed clearly that the steam-era requirement of a separate crew for each locomotive would no longer be strictly applicable. On some roads these disagreements became strongly adversarial, with talk of strikes and other legal actions.

With technology rapidly overtaking work rules, most railroads attempted initially to

Figure 4.5. CB&Q's early success with streamlined passenger trains led to the need for more power on its *Denver Zephyr*. Led by No. 9906 (two 900-hp Winton 201As), with assistance from a booster carrying a 1,200-hp 201A, an 11-car consist slides out of Denver in December 1939. David Price Collection.

sidestep the issue by using numbering schemes similar to UP's, wherein an entire multi-unit set of power carried one primary number with each unit identified by a supplementary letter. Management contended that all interconnected power units constituted one locomotive with a single number. Before this issue was finally settled during a federal lawsuit in 1942 (entitled *Diesel Fireman vs. Western Association of Railway Executives*), some early buyers of diesel locomotives would put neither the railroad name nor any readable number on the side of a booster unit. Others resorted to changing number boards on units between runs in order to make identification sequences consistent.

Passenger Locomotives

Richard Dilworth, EMC's chief engineer, lost no time in getting his company to construct two full-size locomotives that could demonstrate the capabilities of Winton 201A–powered units for hauling conventional passenger trains. In May 1935, two 1,200-hp boxcab units, reminiscent of the Ingersoll-Rand models a decade earlier, left GE's Erie plant for demonstration runs. Heretofore, most diesel-powered trains had been articulated configurations that, with permanent drawbar couplings, were much safer at high speeds (above 80 mph) than conventional couplers of that era. But EMC recognized the vast market for replacing steam with diesel power on conventional trains. Quickly, B&O bought No. 50, a near duplicate of the demonstrators, to power its new *Royal Blue* lightweight train, which was unveiled in April 1936. Eventually both the diesel and its train were transferred to B&O subsidiary Alton Railroad, where it became the *Abraham Lincoln*, while the locomotive was rebuilt with a slight shovelnose contour on one end.

Conspicuously absent during the emergence of the articulated streamlined trains, the Santa Fe purchased two 1,800-hp diesel passenger locomotives in September 1935 to power

Figure 4.6. Principal figures in America's early period of dieselization pose beside Baltimore & Ohio's first EA unit, No. 51, before dedication ceremonies at the EMC plant on May 19, 1937. From the left, they include George Codrington of Winton Engine, Harold Hamilton of ElectroMotive Corp., Charles Kettering of General Motors, and Charles Galloway of B&O. In the lower photo, B&O's No. 52 speeds under a semaphore-laden signal bridge, leading a train out of Jersey City in December 1939. Mark Reutter Collection (top), Robert Malinoski, James Mischke Collection.

Figure 4.7. Differing nose styling of early Union Pacific passenger diesels is evident in this January 1938 portrait of ElectroMotive units in Chicago's North Western Station. From the left, the lineup includes the *City of Denver*, *City of San Francisco* No. 2, and *City of San Francisco* No. 1. Nose herald of E2A at right features the three roads involved in the famous Overland Route (C&NW, UP, and SP). Mark Reutter Collection.

its new *Super Chief* train. Numbered 1A and 1B, they rode on B-trucks and featured a flat face topped by a high-mounted radiator behind a gracefully contoured cowling. Three years later, they were modified from double to single end operation, and their noses were reconfigured to include a turret cab beneath the radiator. The running gear was also changed to a pair of unusual B-trucks with attached guiding wheels (a 1B-B1 arrangement). During the ensuing 15 years, the units would undergo numerous rebuildings before being scrapped in 1952. In 1935–36 Burlington bought the last of its shovelnose units powered by Winton engines. Cars 9904 and 9905 carried a 900-hp V-12 engine, while 9906 and 9907 contained the 1,200-hp V-16 model.

Responses to the three early passenger units (demonstrators 511 and 512 and B&O's No. 50) showed clearly that a brightly painted, streamlined car body was a major part of the diesel locomotive's allure with the traveling public. Thus in 1937 EMC began producing its famous E series of streamlined passenger locomotives. Folklore suggests that the letter *E* stood for "eighteen," indicative of the horsepower produced by a pair of V-12s. Baltimore & Ohio received the first six sets of A-B units (EA and EB models). They featured a long, slanted nose with buried headlight at the top and rode on a new three-axle, two-motor truck (A1A-A1A arrangement) that would become an EMC standard for as long as E-units were produced (until 1964).

During 1937–38 the next E-model (E1) was also produced in A and B car bodies for the Santa Fe, which introduced its famous Indian warbonnet paint scheme and assigned them to *Super Chief* service. In the meantime, Union Pacific ordered the only E2 models (two A-B-B sets) for use on the *City of San Francisco* and *City of Los Angeles*. These units displayed a slightly different variation of the aforementioned numbering scheme, being SF-1, SF-2, and SF-3, along with a similarly designated LA set for the Southern California train. Moreover, the two sets of power units, although painted in UP yellow, were owned jointly by the three roads over which they operated (C&NW and SP for the San Francisco train and C&NW for the LA train). Each lead unit carried nose-mounted heralds for all participating lines.

Figure 4.8. EMC's switcher production was ramped up rapidly before World War II, as this was the quickest and cheapest way for a railroad to ease into the diesel age. Santa Fe's No. 2152 was an SC model powered by a 600-hp, 6-cylinder Winton 201A. Constructed in July 1937, it was retired a few months short of its 50th birthday. It is shown here working the docks at Galveston, Texas, in February 1938. W. H. B. Jones, David Price Collection.

A New Plant and a New Name

With the outstanding successes of its initial passenger locomotives, EMC's Hamilton and Dilworth decided in 1934 to combine car body and truck fabrication with Winton engine construction at a new plant. The company needed to shed its earlier role as an equipment consolidator and become a full-fledged diesel locomotive builder from the wheels to the roof. Although they knew full well that getting funds for a new plant would be a hard sell to the higher-ups at GM, the two EMC men nevertheless showed up at an executive planning conference at C&O's posh Greenbrier Resort in White Sulfur Springs, West Virginia, feeling somewhat out of place.

A typically plainspoken railroader, Dilworth prefaced his request for $1 million by letting the senior executives know how much more important rail transport was to the nation than trucks, which he called "rubber-tired toys." After his presentation, he had little hope of success and thus was quite surprised when Hamilton told him that their request had been granted. But he also learned that they would not get the amount requested and that the location would probably not be in Cleveland, where both EMC and Winton were situated. As it turned out, GM management had allocated $6 million for a new locomotive assembly plant, and soon site selection was begun for a location with easy rail access and a strong bedrock foundation.

On March 27, 1935, the two EMC leaders participated in an official groundbreaking ceremony on a 74-acre slice of cornfield in the Chicago suburb of McCook. It so happened that the post office of the construction company was at nearby La Grange, and that became the official address for EMC's Plant 1, served by the Indiana Harbor Belt Railroad. Although the plant initially fabricated only car bodies, its first complete locomotive was rolled out in May 1936, an SC-model switcher with a 600-hp, eight-cylinder 201A. A

Figure 4.9. The long slanted noses of early EMD passenger units set them apart from all diesel locomotives before or since. Above, Missouri Pacific's first passenger diesels, E3's No.7000 and 7001 (1939), leave Little Rock on a June afternoon in 1960 with a night train for Memphis. CB&Q's stainless-steel E5 A-B units head into Ft. Worth in 1964 with the *Texas Zephyr* (top, facing page), completing an overnight trip from Denver. Finally, a sunny day in 1939 finds the new ACL E6 No. 500 and the *Champion* standing ready to depart from Miami. Note that the retractable coupler is in the outward position (bottom, facing page). J. Parker Lamb, David Price Collection (bottom).

Figure 4.10. EMC offered the 1,800-hp TA model (B-B) to those roads that did not operate high-speed, long-distance passenger trains. Rock Island bought the only six produced for its *Rockets.* Later they were reassigned to short routes such as the *Texas Rocket,* operated by the Burlington–Rock Island Railroad between Dallas and Houston. No. 602 and its three-car train are shown ready to leave the Houston Union Station in 1938. Hugo Lackman, David Price Collection.

contemporary photo shows a companywide celebration with workers posing on, in, and around the boxy Santa Fe unit (No. 2301).

Soon the EMC brain trust decided it was also time to replace the pioneering Model 201A that had been stretched as far as it could reasonably go. The new engine project was begun in the summer of 1936 under the direction of a still young but highly experienced Gene Kettering. With the same thoroughness used in developing the 201A, he and his technical team were able to start afresh in creating the 567-series engine. The final configuration included a 45-degree Vee configuration (vs. 60 in the Winton) with 8.5 × 10 cylinders, a compression ratio of 16:1 (same as 201A), an rpm of 800, and aspiration with a Roots-type blower. Produced in four sizes (6, 8, 12, and 16 cylinders), the 567-model would eventually pass through many upgradings until superseded in 1966 by a larger machine. The number designation of this engine and its successors comes from the piston displacement (cubic inches) of each cylinder.

Intending to develop a line of more powerful passenger locomotives using the new 567 engine and continuing the model sequence E1 and E2, the company produced an E3 demonstrator (No. 822) in March 1939. It was housed in a car body similar to the E1, but it contained a pair of V-12 engines, producing 2,000 hp, which would become the standard configuration until the late 1940s. During the years leading up to World War II (1937–42), EMC produced passenger power in models E3 through E6, all of which were housed in a car body that featured a variation of the slanted nose on the E3 demonstrator. The later models featured an extended headlight housing rather than a recessed one. Together the E3, E4, and E5 models accounted for only 41 A-units and 12 B-units. Seven lines received E3s (including KCS, which bought the rebuilt demonstrator), but only Seaboard bought the E4, and CB&Q bought the E5 model. With 118 units, the E6 model was by far the most popular EMC product before World War II.

Models AA and TA were used by EMC to designate special designs. The first AA was a boxcab passenger unit (B&O No. 50), and the second was a silver shovelnose unit delivered to Burlington in 1939 to power the *General Pershing Zephyr.* It resembled other models of the fleet but carried an asymmetrical wheel arrangement (A1A-2) and a 1,000-hp

V-12 Model 567. This unit is preserved at the Museum of Transport in St. Louis. The TA model was a B-B passenger unit using a V-16 Winton and a shortened E1 car body. Rock Island used six of these on its original *Rocket* passenger trains.

Brute Force for Freight

While flashy streamlined trains brought EMC to the fore as America's leading producer of diesels in the 1930s, one must recall that the inspiration for this giant advance in technology had come from the calculated risks by two large western carriers. But within a few years it was GM's turn to consider a big leap. Buoyed by the performance of the 567 engine in passenger and switching service, the company made a critical and strategic decision. It would offer a direct challenge to a common notion among steam-oriented railroad mechanical departments, namely, that a clean and pretty diesel did not have the muscle for the rough-and-tumble work of hauling heavy freight trains up and down grades. So EMC decided that the only way to win that argument was to demonstrate the performance of a diesel-electric locomotive built especially for freight service.

In 1938 the company began working a B-B freight model powered by a V-16 version of the 567 engine (1,350 hp). Using the same type of car body as passenger units, except for a shorter and blunter nose, the locomotive would include a booster unit permanently coupled by a draw bar, giving a total of 2,700 hp. The model designation was FT, and it has been speculated that the *T* stood for 1,350 hp, just as *E* had designated the 1,800 hp of early passenger units. One of the innovations introduced by the FT was a new truck designed by Martin Blomberg, a Swedish-trained engineer who migrated to Quebec in 1912 and then to the United States in 1916. Blomberg worked for Pullman from 1925 to 1935, where his last assignment was designing the *Denver Zephyr*. Joining EMC in 1935, he was quickly put in charge of designing car bodies and trucks. He led the design team that worked with General Steel Castings (GSC) on the A1A truck for the early E-units.

His primary innovation was the outside swing hanger that supported an elliptical spring-mounted pedestal for the car body. The swing motion allowed the truck to move laterally without disturbing the car body. His work on the two-axle truck represented a more difficult challenge than the three-axle model due to its shorter wheelbase and greater axle loads. Blomberg's final design employed an artistically shaped side frame that was strong enough for the heavy loads of freight service. It became an EMC-EMD icon for as long as B-B freight locomotives were built (ending with the GP60 in 1994).

To showcase its new freight unit, EMC put together in late 1939 two A-B sets of FTs, coupled back to back, producing a machine nearly 200 feet long and weighing over 450 tons. Carrying the number 103 on one end and 103A on the other, the dark green and yellow demonstration units set out on an 11-month national tour in November. Once again, diesel-electric technology startled the railroad world, this time with power and toughness rather than speed and endurance. Many skeptical motive power people had their eyes opened by what transpired before No. 103 returned to the La Grange plant in October 1940 after logging nearly 84,000 miles in 35 states from New England to New Mexico. Tackling grades on lines from Southern Railway to the Northern Pacific, number 103's solid performance, especially in rough terrain, consistently surprised the test engineers and observers making measurements in the dynamometer car that connected the locomotive to its train. Clearly the FT was no paper tiger, and it could (and did) take a big bite out of traditionally slow trips up steep grades normally requiring helpers.

Suddenly EMC, in a repeat of its earlier passenger demonstration scenario, was flooded with potential orders and began to tool up its production lines at La Grange. And it was not a moment too soon, as production during the next four years would eclipse anything Hamilton could imagine. Santa Fe was quickest on the draw, ordering four A-B sets soon after No. 103 returned home. But strong-willed Fred Gurley, AT&SF's vice president, proposed some changes as part of his road's order. Gurley had been with Burlington during the development of the *Zephyr* and was a strong proponent of this new form of power. He told EMC that the new FTs needed to replace the drawbar between units with a normal coupler and include a holding brake on each unit. The latter device, later known as a dynamic brake, functioned by reversing the current flow through the traction motors, turning them into generators whose electric power was dissipated in toasterlike grids mounted on the roof. This action provided a significant drag on the locomotive and thus controlled the speed of downhill operation to an acceptable level. These were commonly used on electric locomotives but had not yet been needed on diesels. As to the coupler between A and B units, this was not as simple a matter as it might appear from a distance. With the dimensions of the drawbar connection deeply embedded in the car body design, it took a good bit of old-fashioned tinkering to come up with a way to get coupler pockets to clear the traction motors at the interface between units, as well as to rewire the units for completely separate operation.

Santa Fe began taking delivery of its order in the final two days of 1940 and had put all units into service by February 1941, a month after EMC was fully merged into GM as its Electro-Motive Division. The second FT order (two A-B sets) went to Great Northern in May and June 1941. These B-units were fitted with small steam generators for possible use on passenger trains. However, before these units could be delivered, GN ordered two more sets (below). Also in mid-1941 Southern Railway received the reconditioned units of the demonstrators set as well as two new A-B sets. In August, the diesel-hungry ATSF received eight B-units that allowed it to assemble four sets of A-B-B-B locomotives, which were operated on long districts where turning would not occur very often. In October GN received its second order, an A-B set and a new A-B-A combination in which the middle unit was shortened and all units were connected with drawbars.

Another A-B-B-B set for ATSF followed in November 1941, and the Rio Grande received six A-B sets between late November and February. With more orders piling up, La Grange was working around the clock to meet demand. It was in the midst of the next batch of orders (ATSF, GN, and B&O) that the frenetic pace was disrupted. On April 4, 1942, the War Production Board (WPB) issued production controls that covered all sectors of American manufacturing, including the major locomotive builders Alco, Baldwin, Lima, EMD, and GE. Although locomotives were desperately needed by the industry, there was a severe shortage of materials, especially copper for electrical gear used in diesels and high-strength steels for steam locomotives. Moreover, locomotive builders were also given orders for government equipment. For example, steam builders constructed artillery pieces and armored vehicles, while EMD built diesel-electric propulsion units for navy ships in early 1943.

The WPB soon began reviewing and allocating locomotive orders to each builder. On the diesel side, it recognized the significant progress made by EMD's production of FT units and thus prohibited any construction of passenger diesels and also assigned Alco and Baldwin to build only switchers and road switchers. EMD received its first allocation letter in May 1942 instituting materials allocations, but it was allowed to continue all types of locomotive production through August. However, only freight power could be produced between September and November. The second letter (November) mandated zero railroad production for December 1942 through February 1943 (navy work) with freight units produced only between March 1 and June 30. Subsequent letters allowed only for FT production. Despite these regulatory restrictions, EMD was able to construct a total of 1,047 freight units during the critical war years (1942–45), peaking at 500 units during 1944. Adding the prewar production (49), the grand total was an astounding 1,096 FT units.

The distribution of FT deliveries is intriguing to analyze, as it indicates roads with high traffic demands (especially movement of petroleum products) while on others it shows a desire to maintain an extensive steam fleet. Of the 23 roads receiving FTs, 19 were major

Figure 4.11. At top, EMC's pioneering FT units, No. 103 and mates, pause for servicing in Denver in the summer of 1940 during their national demonstration tour. Santa Fe bought FTs early and often, building a fleet of 320 units during World War II. In later years, they were mainline workhorses on Texas and Oklahoma lines. Lower photo shows the three-unit set, No. 127, at Temple, Texas, in 1964. Louis A. Marre Collection (top), J. Parker Lamb.

trunk carriers while only 4 were regional independents or subsidiaries. Receiving lines included ten that extended west from Chicago while only three recipients operated from Chicago to the East Coast. In the northeast corner (Pennsylvania through New England) there were five recipients with another five on the southern rim (Virginia to Texas). Not surprisingly, absent as recipients were the steam-powered coal carriers from Pennsylvania southward as well as two of the nation's largest lines (SP and UP), which maintained extensive steam fleets. The two most unlikely recipients were among the last to receive FTs. Minneapolis & St. Louis received four As and two Bs in April 1945, and the New York Ontario & Western (in receivership) took delivery of nine A-B sets in May.

Needless to say, the federal government's intrusion into activities of locomotive builders during World War II was no more popular than the takeover of all class 1 lines by the U.S. Railroad Administration between 1917 and 1920. Many railroad presidents were displeased when their FT orders were delayed for a year or more, while they watched neighboring lines unwrap theirs, and some lines (notably B&O) agreed to settle for modern articulated steamers in lieu of diesels.

As EMD's market dominance grew in the early 1950s, the two largest steam builders (Alco and Baldwin) began to make a legal case against the WPB restrictions, claiming they had suffered irreparable harm by being prevented from developing new diesel units during wartime. These and related matters were explored in hearings by the Senate Subcommittee on Antitrust and Monopoly in 1955. The published transcripts of these proceedings, referenced a number of times in the present narrative, provide insightful data concerning the history of American diesel builders. Nevertheless, the issues involved in GM's rise to dominance during and after World War II led eventually to the federal courts, where in June 1963 the Justice Department filed an antitrust lawsuit. It sought a judgment against GM for monopolistic practices, specifically to divest itself of the Electro-Motive Division, which in 1961 controlled 80 percent of the diesel locomotive market. After years of taking depositions and contesting legal motions, the government filed for dismissal in 1967, citing insufficient evidence. Clearly it was an uphill battle for the Justice Department, since most railroad executives were quite reluctant to discuss possible wrongdoing by one of the most powerful corporations in the world, as well as one of the nation's largest rail freight shippers.

CHAPTER 5

DEVELOPMENTS BEYOND
LA GRANGE

The meteoric rise of EMC-GM as the first major producer of American diesel-electric locomotives tends to overshadow the technical progress made by other builders before World War II. The two leading steam locomotive builders, Alco of Schenectady and Baldwin of Philadelphia, made notable contributions to the evolution of machines powered by IC engines. While both companies teamed with others on various diesel development projects, each produced a sizable number of units under their own names.

Early Baldwin Diesels

This company began its climb to the top echelon of American heavy industry at the dawn of railway development and eventually came to dominate worldwide production of steam locomotives for nearly a century beginning in the 1840s. For the most part, Baldwin's business philosophy was a prudent one: build a variety of machines to suit a variety of customers. President Sam Vauclain was interested in the efficient, but mechanically complex, compound steam engine, but the company seldom risked introducing new technology. However, BLW's executives were fully aware of the diesel engine work by General Electric, Ingersoll-Rand, and Westinghouse in the period surrounding World War I, since Baldwin had constructed car bodies for many WEMCo units.

Taking note of New York's Kaufman Act (1923) banning steam locomotives in Manhattan, the company decided to construct an experimental diesel locomotive in 1925. Eschewing the Westinghouse approach of beginning with a light, low-speed switcher configuration, the proud Vauclain was determined that his company's name be attached only to a mainline machine. What he got from the plant was a double-ended boxcab unit that resembled an electric locomotive, even to its articulated trucks (A1A + A1A). Carrying a road number identical to its serial number (58501), the car body housed a unique engine, an unorthodox inverted V-12 designed by the Knudsen Motor Company of New York. The cylinder banks met at the head while twin crankshafts at the bottom were geared to the generator. Running at 1,200 rpm, the two-stroke machine was designed for 1,000 hp, while its weight of 275,000 pounds was expected to produce 52,200 pounds of tractive effort. Unfortunately, but not unexpectedly, there were many bugs in this experimental machine, and its performance was lackluster at best, due largely to an overweight and unproven

Figure 5.1. New Orleans Public Belt bought the first three Baldwin VO 660s in December 1937 after considering large steam switchers to pull transfer cuts up the 1.25 percent grades on the new Huey P. Long Bridge. A still active No. 33 was recorded in October 1967. W. E. Mims.

diesel that never achieved its target performance level. The unit was returned quietly to Eddystone Works and assigned to plant duty for eight years before being scrapped.

Four years would pass before BLW attempted another experiment in IC engine railroad power. In May 1929, a second 1,000-hp boxcab unit, No. 61000, ventured forth as a heavy-duty switcher (maximum speed 25 mph). The double-ended unit featured a rounded roof contour and again rode on articulated trucks (B+B). This time the power plant was a slow-running (500 rpm) four-cycle Krupp engine from Germany whose cylinder dimensions were a massive 15 inches square (bore and stroke same size). With a weight of 270,000 pounds, all of it on the driving wheels, this model gave a good account of itself in both performance and cost of operation during a demonstration tour of seven class 1s stretching from the nearby Pennsylvania Railroad to the Northern Pacific. However, the company viewed its two diesels as embarrassments because of their numerous breakdowns. It was quite unaccustomed to starting at the bottom of the learning curve.

The modest success of No. 61000 gave some evidence that Baldwin had an improved understanding of diesel locomotive design and, in the eyes of some observers, could have used this experience base to move toward lighter engines and improved car bodies. But that was not to be. Five months after the unit left Philadelphia, the stock market crashed, plunging the nation into economic chaos. By February 1935 the proud and successful Baldwin empire would be forced into reorganization, following thousands of other failed businesses. However, BLW was such a large company that, although production (and income) dropped precipitously during the Great Depression, much of the technical staff continued to plan for the company's future, one that included the diesel engine.

Like others (IR and WEMCo), Baldwin management, in the wake of Vauclain's demise, decided that it needed to build its own diesel and, in 1931, purchased the assets of I. P. Morris & De La Vergne and moved them to Eddystone. The design experience of the De La Vergne engine group and its line of stationary and marine power plants was exactly what

Figure 5.2. Santa Fe's zebra-striped Baldwin VO 1000 (1940) switches streamlined cars at Bakersfield, California, in June 1950. J. Parker Lamb.

BLW needed to move up the ladder as a diesel locomotive builder. The year 1935 saw the emergence of the new VO Model engine, a six-cylinder, four-cycle machine that produced 660 hp at 600 rpm. A year later, the first VO was lowered onto the cast main frame of No. 6200, an end-cab B-B switcher for demonstration use.

After a year of road testing, the 6200 became Baldwin's first commercial sale when it was transferred in April 1937 to Santa Fe in Chicago as their No. 2200. In January 1937, the New Orleans Public Belt Railroad signed on for three 900-hp models that were delivered in December. The Public Belt found a pair of these units to be perfect for lugging long drags up and over the new Huey P. Long Bridge crossing the Mississippi River. But soon the VO engines began having problems, largely due to the stiffness of their cast blocks, which cracked instead of bending slightly when on rough track.

In 1938 the troublesome engine blocks were replaced by welded ones while General Steel Castings was tapped to supply new main frames. The engine was completely reworked with new crankshaft bearings, cylinder heads, fuel injectors, and an improved generator connection. Santa Fe's Fred Gurley also analyzed their 660-hp unit and provided engineering advice on truck design. Using these improvements, the company decided to launch a new line of switchers, called VO660 and VO1000, to be powered by Westinghouse electrical gear. In April 1939, BLW No. 229, the first VO660, was rolled out for display in a light blue paint scheme with gold lettering. It was followed by a pair of VO1000 demonstrators in 1940. Baldwin had finally assumed a position in the diesel locomotive market.

The first of the larger switchers were ordered by Missouri Pacific (one unit in April 1939) and Santa Fe (five in June 1939). All six were delivered in November, and this motivated BLW to begin constructing stock units. By late 1940, four roads (Reading, Milwaukee, Central of Georgia, and EJ&E) had placed orders for both sizes of units. Six additional lines ordered only the smaller size, while four others chose only VO1000s, so that 15

VO660s and 22 VO1000s were in service by the end of 1940. The following year was even better, with orders for 42 small units (19 lines) and 52 large ones (also 19 roads).

After the WPB allocations began in 1942, the company's locomotive production (of steam and diesels) became intermixed with other wartime needs (including scores of large ship propellers or screws). Output for 1942 was 42 VO660s and 49 VO1000s, including Pennsy's first diesel power (four small and six large switchers). Total BLW production through the wartime period (1937–45) was 143 small units and 548 large ones, including 66 VO1000s for the army and navy. Among the largest owners of Baldwin VO switchers were ATSF (59), Frisco (38), CB&Q (30), NP (28), SP (25), and B&O (25).

It is clear from the foregoing summary that Baldwin had great difficulty transitioning into the broader technology base demanded by the diesel-electric locomotive. This suggests that its basic manufacturing methodology was unchanged during the initial phase of diesel production. Due to its long experience as a heavy machinery builder, it was wedded to the steam era notion that each new locomotive design, based largely on a specific railway's needs, was a separate project that would lead to a limited production run. Early on, there was little consideration of developing standardized components that could be used interchangeably between many designs. Moreover, BLW's electrical systems were unreliable, and this proved to be its Achilles' heel.

Early Alco Diesels

As noted, Alco sneaked into the diesel era in 1924 when its name, along with IR and GE, appeared on the eclectic builder's plate for the first IR boxcab switcher, even though the company had contributed nothing to the car's fabrication. Of course, it later built 42 electric locomotive car bodies for GE and 33 for the GE-IR-Alco consortium before it ended in 1937. Moreover, hidden within the labyrinth of ancillary programs at Alco, events were quietly unfolding for its later emergence as a strong rival to industry giant EMD. The scenario began in 1926 when the New York Central asked Alco, its primary supplier of steam locomotives, to work with another New York company, McIntosh & Seymour of Auburn, a major builder of diesels for stationary and marine use.

Founded in 1886 by John E. McIntosh and James L. Seymour, the plant of McIntosh, Seymour & Company first constructed stationary steam engines. Although their products included no technical defects and were generally reliable, the market for such power sources disappeared around 1910, and so the company turned to diesel engines. In 1913 they entered into a licensing agreement with the Atlas Diesel Engine Company of Stockholm, after negotiations that involved meetings with Rudolf Diesel himself. Upon the retirement of McIntosh, who gave his interest in the business to his partners, the company was renamed McIntosh & Seymour Company, and it soon produced its first engine based on the Atlas design. However, by 1917 M&S had replaced much of the Atlas design with one having more compatibility with American production methods. By the end of World War I, M&S was the world's largest company devoted solely to the manufacture of diesel engines, and it built the largest four-cycle engines in the United States.

In 1925 the company decided to test the locomotive market and produced two prototype engines in V-8 and V-12 configurations using 8 × 9.5 cylinders. These engines, as had all earlier M&S machines, used air injection wherein both high-pressure air and fuel were injected separately, rather than solid injection in which only fuel droplets were injected into the cylinder. The air injection technique was more complex mechanically and eventually was discarded by most engine builders. The first V-8 (200 hp) was installed in a New York Central motor car while the 300-hp V-12 went to the Brill plant in Philadelphia for installation in an M&S boxcab B-B locomotive, which was sold to the Lehigh Valley in early 1927. Performance of the engine was generally poor (due mainly to the injection system), and the unit was soon retired to storage.

Figure 5.3. In 1939 L&N bought both Alco and EMC switchers for comparative tests. The Alco entry, a delicately striped HH 660 No. 10, works the New Orleans yard in September 1949. E. M. Kahn, J. G. Lachaussee Collection.

NYC was the buyer of the next locomotive, and it was here that Alco was called in to improve an earlier IR locomotive design (2-D-2) produced by the GE-IR-Alco consortium. Once again the large M&S engine (80 pounds per horsepower with 14 × 18 cylinders) was a poor performer, and there were no buyers on the horizon for future M&S–powered locomotives. However, Alco's management soon realized there was considerable value in the diesel knowledge and production facilities of M&S as part of Alco's future diesel efforts. Thus in 1927 the Schenectady company began buying its stock, and by December 1929 it had acquired 97 percent of M&S, which could then be operated as a subsidiary. Its first goal under Alco was to construct a locomotive-compatible engine, which occurred in October 1930. The 300-hp Model 330 was a six-cylinder, four-cycle, 900-rpm machine that used solid injection.

In January 1931 the new engine was installed in boxcab demonstrator unit No. 300, Alco's first diesel locomotive as an independent builder. It was soon sold to the Jay Street Connecting Railroad in Brooklyn. However, with the end-cab configuration becoming standard for switching locomotives, the next demo unit was of this type. It was promptly sold to Lehigh Valley as No. 102. Indeed, the performance of the new engine was so good that the railroad brought the unsuccessful M&S-powered boxcab unit out of storage and sent it to Auburn for reengineering with the new Alco diesel.

Alco's first 600-hp engine, the six-cylinder Model 531, was installed in demonstrator No. 600 in 1931. It featured a higher hood to accommodate the larger engine, and after successfully demonstrating on the New Haven, it became their 0900. However, the first of what would be Alco's standard switcher configuration was No. 601, completed in July 1932. Its hood was almost as high as the cab roof, and it rode on Alco's patented Blunt-trucks, conceived by designer James G. Blunt in 1922. They would be used on Alco switchers until the 1950s, when the Association of American Railroads switcher truck (AAR Type A) was introduced. The second 600-hp unit was completed in 1932 and became Lackawanna No. 401. It included a welded main frame in lieu of a large casting (for cost reduction). Soon the road ordered seven duplicates for 1934 delivery. The final of the planned three demonstrators (No. 602) was then built and sold to the Boston & Maine in late 1934.

These high-hood switchers reached their final external configuration in April 1934, when a stock unit was completed. It included cosmetic improvements carried out by styling consultant Otto Kuhler, who cleaned up the lower cab area, gave the hood more rounded corners, and faired its top as a continuation of the cab roof contour. Designated internally as Model 404-DL-132, it, along with its high-hooded siblings starting with No. 600, was commonly designated as the HH600 model by later chroniclers of diesel motive power history. The first of the Kuhler-styled units went to the Belt Railway of Chicago as No. 300.

By 1935 Alco had constructed two dozen of the HH600s, but with rival EMC equipping its switchers with 900-hp Winton engines, Alco engineers decided to increase their power level to meet the competition. The result of their study led to a major technological jump in engine design. As noted in earlier discussion of IC engine fundamentals, one can increase power by increasing displacement or by augmenting the aspiration process. The company decided to compare costs of both approaches. It built an eight-cylinder version of its Model 531 for testing, while simultaneously investigating the Buchi design for turbocharging the six-cylinder engine.

Swiss engineer Alfred J. Buchi had perfected a turbine-driven supercharger during the 1911–14 period under the sponsorship of Sulzer Brothers. After receiving U.S. and German patents, he established a syndicate for granting manufacturing licenses to companies in Europe, among which was the Swiss firm Brown-Boveri Ltd. After initial contacts with Buchi suggested that supercharging would provide as much as a 50 percent power increase with only a 15 percent weight increase, Alco decided to pursue a manufacturing license. This was received in July 1935, just a few months following that of another engine builder destined to play a role in American dieselization, Cooper-Bessemer Corporation. Implementation and testing of the first turbocharged engine, designated as Model 531T, was carried out in May and June 1936 and resulted in an order to Brown-Boveri for a dozen supercharger units before the end of the year. Although typical turbine failures would occur during the first months of railroad service, eventually these were rectified, and the accessories became quite reliable. Since the turbines ran at very high rpm (10,000) and were required to withstand hot exhaust gas, most early problems involved installation of better ball bearings as well as a more effective water cooling system for the turbine jacket.

To increase engine power in 1938, the Auburn plant increased rpm from 700 to 740 and released the new Model 538 engine that produced 660 hp with normal aspiration and 1,000 hp with supercharging. In addition, with war clouds hovering over Europe, Alco began building the Buchi superchargers itself. By 1940 the Elliot Manufacturing Company of Pennsylvania obtained a license from Buchi and became the major U.S. supplier (including Alco). Total production of HH model switchers (600, 660, 900, and 1,000 hp) between 1931 and 1940 included 81 powered by the Model 531 engine and 96 by the Model 538. Of the 177 total, turbochargers were installed on 55 units.

The next Alco design change was also motivated by comparison with models being produced by Baldwin and EMC. Alco's high engine-hoods, which resulting from laying the engines on the upper surface of the main frame, resulted in diminished crew visibility in comparison with other models. Thus in early 1940, design changes were made in the engine mounting lugs and the oil pan shape so that the lower surface of the diesel was some 16 inches below the floor of the main frame. Additional changes allowed the hood height to be decreased by 27 inches. The modified engines were designated as Models 539 and 539T. These new locomotives would carry the external designations S-1 (660 hp) and S-2 (1,000 hp) and would be Alco's mainstay under WPB-controlled production. Total production between 1942 and 1944 was 134 S-1s and 499 S-2s.

In 1934, soon after the appearance of the Union Pacific and Burlington streamlined trains, ACF was approached by Isaac Tigrett, president of the Gulf, Mobile & Northern,

Figure 5.4. Nashville Chattanooga & St. Louis's red-and-yellow Alco S-1 No. 4 (1941), one of the road's first diesels, is busy in Jackson, Tennessee, in September 1954. J. Parker Lamb.

about a new diesel-powered, streamlined train in which the cars would have traditional couplers so that the train consist could be changed. The two companies had been doing business for years as the railroad built up a sizable fleet of gas-electric motor cars. Their final design, generated by Otto Kuhler, was similar to the shovelnose *Zephyr* trains, with a buffet-coach and a sleeper-observation car attached to the power unit, which also contained a baggage-RPO section and was powered by Alco's 600-hp Model 539 engine operating a rear power truck (2-B arrangement). Three power cars were constructed (two in 1935 and one in 1937) to cover a split-route that ran southward from Jackson, Tennessee, with separate sections in central Mississippi to reach New Orleans and Mobile. The GM&N *Rebels* were the South's first streamlined, air-conditioned trains and proved to be highly reliable and financially successful. This operating experience convinced Tigrett that diesel locomotives were exactly what were needed by his railroad.

Construction of EMC's McCook plant in 1935 and its plans for a standardized line of passenger and freight power were major challenges to Alco and motivated the company to think about significant expansion. Its search for a strong electrical supplier led to GE, which was then specializing in small diesel units. In 1940, the two companies agreed to market mainline locomotives under the name Alco-GE. With the passenger power market soaring, the first unit was to be a competitor for EMC's E series. Styled again by Kuhler, the DL-100 Model series displayed a distinctive sculpted nose treatment accentuated with dual headlight housings. Powered by two 1,000-hp 538T engines, it also introduced the Commonwealth A1A truck, which became an Alco standard for passenger locomotives.

The first production model, a DL-103, went to Rock Island in January 1940, and eight months later a pair of slightly improved DL-105s, with virtually the same car body but powered by 539T engines, went to Isaac Tigrett's new railroad, the Gulf Mobile & Ohio, for its St. Louis–Mobile *Gulf Coast Rebel*. Rock Island received the last DL-105 in November 1940 and the first DL-107 in December. In 1942 production grew steadily as eight DL-107 models went to Santa Fe, Southern, C&NW, Milwaukee Road, and Rock Island.

Figure 5.5. Alco's DL series of passenger diesels were styled by Otto Kuhler to be distinctly different from EMC's units. Top photo shows GM&O No. 270, the first DL 105, delivered in 1940 for use on the *Gulf Coast Rebel*. On its daily run from Mobile to St. Louis in June 1955, the train stops at dusk in Meridian, Mississippi, for water and a crew change. At bottom, Southern Railway received two Alco DL 107 A-units in February 1941 for use on the *Ponce de Leon* between Cincinnati and Jacksonville, Florida. The two are shown at Atlanta in 1942. J. Parker Lamb (top), Shelby Lowe, J. Parker Lamb Collection.

Production in 1942 (until halted by WPB) included 3 DL-108s to Santa Fe and Southern, 20 DL-109s to New Haven, and a DL-109 to GM&O. Only one B-unit (a DL-110 for Southern) was built during this period. Total prewar production of the DL series was 38 units. However, Alco and New Haven were able to convince the WPB that the DL-109s were dual-purpose locomotives, and thus NH was able to accumulate a roster of 60 such machines (88 percent of total production). With diesel-electric technology, variable gear ratios between traction motors and axles produced this additional flexibility not available with steam power. Unfortunately, the overall performance of the DL models was less than ideal, due largely to the more rigorous demands on the 539 model engines in long-distance service rather than in low-speed, start-stop yard duty.

Although a few roads began dieselization with an all-Alco fleet, many others used both Alco and EMC power in interchangeable roles. On those lines where Schenectady units were a distinct minority, they were often concentrated in specific regions and shops, in which Alco-trained technicians maintained them.

General Electric's Second Generation

General Electric had made pioneering contributions through its Gas Engine Department, and later its Locomotive and Car Department was a big supplier of electrical components to other builders. However, the conglomerate also took great advantage of a provision in the 1937 railroad labor agreement that allowed one-man operation of any common-carrier locomotive weighing less than 90,000 pounds. In response, GE had developed a succession of popular industrial and shortline locomotives, known simply as GE 44-toners. This diminutive machine, a center-cab unit only 33.5 feet in length but powered by two 200-hp diesels, was almost as versatile as a heavy truck and was bought by railroads of all sizes. For example, the owner of the largest number of 44-toners was none other than the Pennsy with a roster of 45. During an unusually long production run of 16 years, beginning in 1940, these units were constructed in five variants, each with slightly different appearance details. A total of 359 were built for U.S. lines with 9 for Canada and 5 for Mexico. Between 1946 and 1956, GE also built two end-cab models having nominal weights of 70 and 95 tons and a gross power of 600–660 hp. Total production of these units was 239 (U.S.), 38 (Canada), and 8 (Mexico).

A Significant Innovation

A second major contribution to diesel locomotive evolution by Alco had its beginning in 1940 when Rock Island president J. D. Farrington, having just put into service the first DL-103, met with company representatives in his Chicago office. During the meeting he sketched out his ideas regarding a combination locomotive that could work in the yard as well as on branch lines in low-speed passenger and freight service. The concept resonated with the Schenectady group, and soon they proposed what became known as a road switcher. Diesel locomotives up until this time had been developed in an industry that still carried a steam era mind-set. This meant that a different locomotive design was required for each type of service: freight, passenger, and yard. The recognition that a diesel-electric locomotive was not bound by these rigid service definitions represented a major conceptual breakthrough that would eventually reshape attitudes about motive power utilization within the entire railroad industry.

To achieve the first road switcher, later designated as RS-1, Alco designers placed an S-2 switcher body atop an extended frame (8.5 feet longer) that rode on General Steel Castings Type B trucks (longer wheelbase than Type A). The primary visual characteristic of this unit was a short hood over the extended platform. The hood was intended to house a steam

Figure 5.6. General Electric's 44-ton and 70-ton units went mainly to short lines and industries, but some toiled for mainline roads. At top is Frisco No. 8, built in December 1943 for a subsidiary, Kansas City, Ft. Scott & Memphis. When photographed at York, Alabama, in 1954, it was assigned to the Alabama Tennessee & Northern subsidiary, where it ran until 1972. In 1958, the Norfolk Southern Railway bought three 70-tonners for use on its light-rail branch to Bayboro, North Carolina. At bottom, a pair of the red units scurries toward New Bern in May 1962 with a train of wood chip hoppers. Unlike most small GE's, these were fitted with MU-connections for heavier trains. J. Parker Lamb.

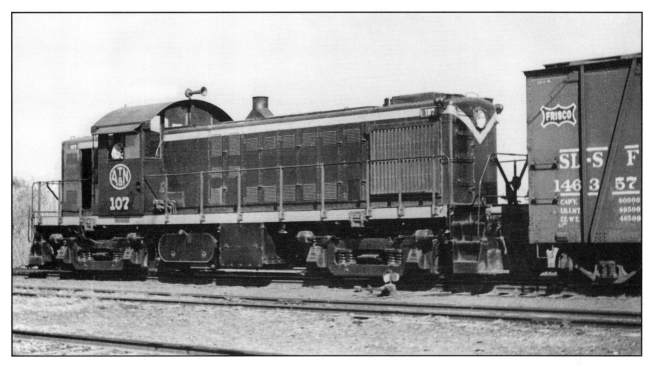

Figure 5.7. Alco's RS-1 represented the first attempt to design a diesel locomotive with the capability for both mainline and yard duty. At top, Alabama, Tennessee and Northern No. 107 (May 1946) switches at the Mobile yard in May 1950. Below, GM&O No. 1105 (1944) displayed its final configuration, which included roller-bearing trucks, bright red paint, and a silver cab roof as it worked the Montgomery, Alabama, yard in July 1967. J. Parker Lamb (top), Ed Mims.

generator for passenger service or additional fuel capacity for freight work. Rock Island quickly ordered two of these units for tests, followed closely by a pair for the southern short line, Atlanta & St. Andrews Bay. Since this design fell within the WPB-imposed upper limit of 1,000 hp, a fleet of 13 units was constructed by mid-1942 for the following roads: Rock Island (4), A&SAB (3), and two each for Milwaukee Road, New York, Susquehanna & Western, and a U.S. Steel line in Birmingham, Alabama (Tennessee Coal & Iron).

What happened next illustrates the extreme measures that were necessary during the early phases of World War II. In this case, there was a crisis involving the 900-mile Trans-Iranian Railway, which represented a key supply route to Soviet Russia, then in a desperate struggle with German armies near its industrial heartland. The need was for a fleet of relatively simple diesel locomotives that would allow a quadrupling of the supply tonnage over this hazardous route from the Persian Gulf to the Caspian Sea. The rail line would be operated by trained U.S. army personnel.

A series of meetings involving the War Department and the WPB produced an agreement that using the versatile RS-1, already proving its mettle on American lines, would be the quickest solution, since Alco could do the assembly on a fleet basis. However, the final solution was something of a surprise. The War Department requisitioned the entire 13-unit group of RS-1s already in service and ordered Alco to build 44 new ones, 9 of which were already under construction. The conversion to foreign service was not a simple matter, largely due to the poor condition of the railway itself. Thus a new C-truck was designed that lowered axle-loads and, due to its extreme flexibility, could navigate rough trackage. The new model was designated RSD-1 and, besides its trucks, was characterized by a tapered cab for clearance inside tunnels.

Alco completed 34 new units by the end of December 1942, 9 in January, and 14 in February. Due to the Nazi submarine threat in the North Atlantic, the units were shipped around Africa to the Persian Gulf, and rail operations began in March 1943. After this initial crisis passed, the War Production Board allowed the company to replace all but two of the requisitioned units in April and May 1943. The only exceptions were for the steel company line in Alabama, which didn't get its units until after the war (1946).

But Alco's road switcher production continued to thrive. In 1944 there was a military order for an additional 100 units, of which 70 were destined for the broad-gauge (5-foot) Soviet railway system. They were well received by the Russians and widely duplicated after the war, an estimated 6,000 being constructed for USSR rail network. The remaining 30 military RSD-1s were used by the U.S. army on lines in France and Belgium before being returned to the United States. After the war, most of these units, now declared surplus, were spread around to other government agencies, including the government-owned Alaska Railroad.

Even though Alco produced more powerful road switchers after World War II, there was still a trickle of RS-1 production to U.S. shortlines as well as to the overseas market. The last of 357 RS-1s were a pair delivered to the Grand Trunk Western in October 1957, making this pioneering design the longest continuously produced U.S. diesel locomotive.

CHAPTER 6

ALCO REBOUNDS

Wartime manufacturing restrictions were lifted on January 5, 1945, although the Pacific war continued for seven more months. Rail equipment companies, and especially diesel locomotive builders, anticipated the appearance of a gold rush mentality among railroad executives. For four long years, the nation's rail lines had successfully coped with traffic levels far beyond their design capacities. Now they faced the daunting task of rebuilding beaten-down roadways and worn-out rolling stock. This situation was amplified by the presence of nearly 39,000 steam locomotives, some built during the war, whose final destiny was already determined by the outstanding performance of wartime diesel-electric power in yard, freight, and passenger service. There was virtually no doubt throughout the industry that the diesel switcher had rewritten the book on economics of yard operations, illustrated clearly by some 2,200 switchers that had been constructed by the end of 1946. Alco led the way with over 1,040 units, with Baldwin a respectable second at nearly 700, and EMD, which had begun production in 1936 but was later limited by the War Production Board (WPB), was third with around 500.

Of course, anticipation of the ultimate victory of mainline diesel power over steam was not yet unanimous, especially among those railroads with strong and traditional ties to the coal industry in Pennsylvania and the Pocahontas region (such as PRR, C&O, N&W, and Virginian). In addition, there were other large roads with an enormous investment in almost new steam locomotives and their related shops and, in some cases, steam fabrication facilities. These roads were proud of the notable performances of the so-called Super Power designs, and their executives still anticipated that a more modern steam locomotive would emerge during the postwar rebound to compete with the diesel on the main line.

The experiences of the Pennsylvania Railroad show clearly the price of continuing with steam well after mainline diesel power had proved itself. During the mid-1940s, when EMC began selling hundreds of E-series and F-series diesels across the nation, Pennsy was spending millions to acquire a fleet of steam locomotives using a design known as duplex drive. To reduce the sizes and weights of the large running gear needed for two-cylinder locomotives such as 4-8-4s and 2-10-4s, a second set of cylinders was spliced onto the rigid frame to produce much lighter rotating masses. Final models included a high-speed passenger engine, the T1 4-4-4-4, and a powerful freight locomotive, the Q2 4-4-6-4. Even though these were among the most sophisticated and efficient steam machines ever built, they still suffered from the inherent shakedown maladies of experimental machines that were rushed into production. Moreover, PRR remained saddled with the steam-related

Figure 6.1. Isaac Tigrett, GM&O's president, was first in line to buy postwar Alco freight units and completely dieselize his road. FA-1 No. 704 was one of the first units leaving the plant in May 1946. Four years later, it leads an FB-1 mate on a northbound train leaving Meridian, Mississippi. J. Parker Lamb.

costs of fuel and water supplies, servicing facilities, and maintenance expense. By 1945 the nation's largest railroad was hemorrhaging red ink due to its rigid view about steam power.

To get a clearer picture of the cost of this sentimental stance, PRR president Martin Clement moved James M. Symes into the post of deputy vice president of operations in October 1946. During his previous position in Chicago, he had become quite familiar with dieselized operations on the Burlington and the Minneapolis & St. Louis. Thus his subsequent study of diesel economics provided objective results that were shocking to many stubborn steam supporters. One of his first reports in 1947 compared the costs of moving freight with Q-2s and with contemporary diesels throughout the western end of the railroad. Including all ancillary costs, the result was a $9.9 million annual loss. Further study showed that the use of two or three steamers on long coal trains was especially costly. On the hilly Renova Division in western Pennsylvania, annual costs saved by replacing steamers with 6,000-hp diesels would be about $1.5 million alone.

Based on this incontrovertible evidence, Pennsy jumped into dieselization in a big way, starting in 1946, by purchasing moderate numbers from every builder and fleets from some. The road went from 34 switchers in 1947 to 549 units at the end of 1949. Of these, 145 were for mainline trains. Moreover, an additional 226 units were on order. But there was also a downside for the territory Pennsy served, involving the human cost in jobs and decimation of local economies. When the famous Altoona shops complex was closed, 3,500 jobs were lost. Moreover, related companies also felt the sting. Baldwin Locomotive Works, a Philadelphia neighbor of the Pennsy since its inception, had been a major supplier of PRR steam locomotives. It immediately lost the steady flow of PRR steam orders, and some years later, when BLW's diesels were not competitive for PRR orders, the loss of its business was a major reason for its demise (see chapter 7).

In other sections of the nation, diesel acceptance came much more quickly. For example, many southern rail lines and the midwestern farm belt were built with light roadbeds that limited axle loadings. Moreover, near coastal regions, other routes were lined with weight-restricted timber trestles. In these regions, railroad executives were convinced that the diesel's heroic wartime performances were not a fluke and that its inherent flexibility would allow them to assemble powerful locomotives with light-axle loads, merely by adding units. They were ready to buy more diesels as quickly as they could line up the financing.

As soon as it was apparent that government controls would be lifted, designs were finalized and production plans were begun at the three diesel builders, Alco, Baldwin, and EMD. While previous discussions have followed the separate developmental periods of these companies, this chapter is directed to the postwar emergence of Schenectady as the primary competition for industry leader EMD. Even though Alco's prewar diesels (539 model) were only moderately successful, by early 1940 the company's engineering staff had begun working on the Model 241 as a competitor for EMD's 567 series. Engine development work was expanded to both the Auburn (old M&S) and Schenectady plants, and many new materials were tested for high-stress components, including improved superchargers.

In 1943, WPB gave its construction approval of an A-B-A set of experimental locomotives powered by the 241 engine. However, due to wartime production schedules, the three units did not emerge from the plant until September 1945. Painted solid black, they were dubbed "Black Marias," and they ran tests with little public notice on the Delaware & Hudson, New Haven, and Bangor & Aroostook. In October they raised some eyebrows on the BAR by hauling a 205-car train. But this significant event proved to be their last gasp. Serious engine failures soon caused the test engineers to tow them back to Schenectady, where they were dismantled and used for spare parts.

In early 1944, Alco's Diesel Engine Department began work on an improved machine, Model 244. It had a 45-degree Vee-alignment similar to the 241, but unlike that engine, it used cooling water only for the cylinder liners rather than for the entire engine block. With 9 × 10.5 cylinders, the engine ran at 1,000 rpm, was aspirated by the latest GE turbocharger, and also included a new drop-forged crankshaft from Erie Forge Company. Conversion of former gun shops at Schenectady into engine development facilities began in late 1944. They produced the first V-16 (2,000 hp) 244 models in early 1946. In the meantime, Auburn had produced 35 Model 244s (V-12 and 1,500 hp) that were ready for installation as early as October 1945. Although there would be the usual teething problems, the Model 244 finally had propelled Alco into a competitive position with EMD.

Evidence suggests that Alco's continuing difficulties in getting its postwar power plant into final production was compounded by internal friction between the engineering staff that led the 241 project and those managing the later 244 program. The company suppressed such discussion to the extent that it could, so that very little formal information is available. The crux of the disagreement centered on whether to make incremental improvements to the 241 design or to start anew on the 244 model and have to climb up the learning curve again. Obviously those in the latter group held more influence in the executive suite. In hindsight, the six-year gestation period of Alco's locomotive diesel engine shows that, irrespective of wartime exigencies, designing and building a high-performance diesel engine remained a challenge even with the presence of some 3,500 units on American rails. However, in any comparison of EMD's 567 to Alco's 244, one must consider the level of technology involved. The former was a relatively simple two-cycle machine with blower-produced aspiration (running at 800 rpm), whereas the Alco machine was a rather complex four-cycle engine running at 1,000 rpm and driving a high-pressure turbocharger.

In parallel with the engine work was a car body program carried out under the previously mentioned 1940 joint-marketing agreement between Alco and General Electric. The design program began in early 1945 under the direction of Richard Patten of GE's Appearance Design Division. Based on his experience in the automotive industry, the team developed sketches, followed by clay or wooden models for initial reviews. A full-scale clay mockup was then constructed on the shop floor so that the necessary sheet-metal templates could be developed. Their goal was to create a locomotive that appeared to be powerful, fast, and striking (i.e., attention getting). Clearly the passenger unit was to be Alco's flagship, but key appearance features would be carried over to the companion freight unit. The focal point, of course, was the nose, which was long and graceful but featured a flat face, emphasizing a wraparound configuration. A fluted, square headlight

Figure 6.2. Because of its proximity to Alco's plant, the New Haven was naturally a strong customer of Schenectady units. A five-unit set of FA/FB units charge down the main with a manifest near Walder, New York, en route from Maybrook yard to Poughkeepsie in June 1963. Jim Shaughnessy.

mounting also gave the nose a point of accentuation. The unit's forward side-grills were decorated with a curved molding, which also served as a rain gutter over the cab door, while a curved fairing extended the body contours to cover the underbody tanks. Overall, the unit was considered by many observers to be the most visually attractive diesel unit ever built. Indeed, it is the only one to have been repatriated from a foreign country for rebuilding after an extended period of deterioration.

Alco's plan was to introduce both passenger and freight units in early 1946. The fast racer rode on GSC Commonwealth A1A trucks and was designated as the PA-1 (V-16 Model 244) while the shorter, but similar appearing, freight unit was the FA-1 (V-12), and it rode on Type-B trucks. Corresponding boosters (PB/FB) were also to be marketed. Due to a shortage of V-16 engines for the passenger units, an A-B-A set of freight units was the first to roll out of Schenectady (January 4, 1946). Painted in black with light trim, they performed impressively in their road tests on the Delaware & Hudson, accumulating 13,400 miles in 46 days and easily lugging tonnage over grades from 0.8 to 1.4 percent.

But then a bolt from the blue caused all work to cease at both Alco plants. Two weeks into the D&H tests, a nationwide steelworkers strike shut down all production until the last week of March at Schenectady and two weeks later at Auburn. Thus it was May before production and testing programs were back to full speed. The immediate task was to fulfill GM&O's standing order. This railroad, created by Isaac Tigrett from a merger in early 1940, was quite pleased with its previous Alco diesels (S-1, S-2, RS-1, DL-105/109, and the ACF *Rebel* streamliners), and it had an open order to purchase Alco freight units as soon as WPB restrictions were lifted.

The three demonstrator units, now adorned in GM&O livery, were rolled out in May

Figure 6.3. At top, a nearly new set of Erie FA/FB-2 units blast through Marion, Ohio, with an eastbound manifest on a fall day in 1955, blowing past a famous NYC Hudson waiting at the passenger station. At bottom, L&N FA-2 No. 357 (1952) leads two mates and a southbound coal train near Hubers, Kentucky. J. Parker Lamb (top), J. G. Lachaussee Collection.

1946. During the next nine months, the plant produced 55 As and 22 Bs, all lettered for GM&O. The 77 units represented a giant step in Tigrett's determination for this southern line to be the nation's first class 1 road to fully dieselize, which it did in 1949. New Haven, another steady Alco customer, received 15 A-B-A sets in May 1947 to completely dieselize its important Maybrook line in New York State. Among other major roads, New York Central and Union Pacific acquired big blocks of FA/FB units in the late 1940s. Total production of this design between 1946 and 1951 was 417 As and 229 Bs, plus an additional 48 units built in Montreal.

By 1950 the original designs were replaced by the FA-2 and FB-2 models, which were two feet longer to better accommodate a small steam generator for passenger service, as well as an improved Model 244D engine with 1,600 hp. The largest fleet of these models were the 80 As and 50 Bs on the New York Central while the Missouri Pacific was second with 60 As and 42 Bs. Total production of this series was 369 A-units and 192 B-units.

Delayed by the strike, the graceful passenger units finally emerged from the shop on June 26, 1946, as A-B demonstrators painted black. After a month of tests on the Lehigh Valley between Newark and Buffalo, the demo units, already destined for Santa Fe, returned to the shop in preparation for their formal introduction. On September 18, the sun was shining brightly as a 6,000-hp A-B-A set, resplendent in the flaming warbonnet paint scheme and their official numbers 51L-51A-51B, broke through a ceremonial paper banner at the end of the finishing shop. Alco covered all bases in the ceremonies, even to attaching a special builder's plate carrying the number 75000. Although not technically correct (actually 74696), no one was concerned during this cheerful moment in Alco's history.

A large crowd of Alco, GE, ATSF, and shop union representatives listened to master of ceremonies Percy Egbert (a future Alco president) extol the earning power and operational capabilities of the new locomotives, denoted No. 6000 for their total horsepower. His remarks emphasized the performance highlights of the road tests on the LV. Following the rollout ceremonies, the units were delivered by New York Central to Grand Central Station, specifically to the private siding used by guests of the Waldorf Astoria Hotel. Thus began Alco's five-day round of formal receptions and dinners attended by important personages in the railroad industry. It was equivalent to the coming-out party for a debutante. At the end of the celebration, the pride of Schenectady was delivered again to the NYC for transport to Chicago and the beginning of Santa Fe's demonstration tours.

Later in 1946 another A-B-A set was delivered to Santa Fe, and two A-units went to GM&O. In early 1947, two A-B-A sets were delivered to Rio Grande, followed by units to Union Pacific, Pennsy, and Nickel Plate, while first-time buyers the following year included NYC, Lehigh Valley, Southern Pacific, and the Pittsburgh & Lake Erie (NYC). The year 1949 saw the last PA/PB orders as Erie, P&LE, Katy, Wabash, and Missouri Pacific became owners. The production run of 199 (161 As and 38 Bs) was only about 40 percent of the sales by EMD's postwar E-7 model that sold 510 during the same period. Of course, that was an inherent risk in pitting a newly developed machine against an upgraded version of one that had been produced for seven years.

Although this result was doubtless a great disappointment to Alco's management, there were many others who had a different opinion. Those who observed, admired, or wrote about passenger trains were almost universal in their admiration of the PA's styling. One early manifestation of this recognition occurred in early 1947 when the American Heritage Foundation was formed to operate a nationwide rail tour of historical documents relating to the founding of the United States. A group of heavyweight baggage and sleeping cars were rebuilt by Pennsy's Wilmington shops to include special protective enclosures. The cars were not only crash proof but also included special controls to preserve the historic documents under precise temperature and humidity conditions. The power was supplied by a PA-1, No. 1776, painted white with red and blue stripes to match the train's seven-car consist. Beginning with a three-day exhibit in Philadelphia on September 15–17, 1947, the train stopped in 313 cities in 381 days, ending in Havre de Grace, Maryland, on October 26,

Figure 6.4. Santa Fe received the first PA models with great fanfare in September 1948. Unfortunately, their disappointing performance shoved them to the less glamorous runs, exemplified by No. 75 heading a local at Wasco (near Bakersfield, California) in June 1950. J. Parker Lamb.

1948. Late requests from other cities extended the tour through additional 13 stops, concluding during Inaugural Week at the nation's capital in January 1949. Over the previous 16 months, the PA's graceful and striking exterior had been seen by 3.5 million people. After returning to its birthplace, the 1776 received a metal plaque of recognition below the engineer's cab and became GM&O No. 292, working out its remaining years on the *Gulf Coast Rebel.*

Following the upgrade of the FA-1 to FA-2 in 1950, the revamped Model 244D engine (2250 hp) was applied to the PA-2 and PB-2 models. This engine upgrade was largely a matter of increasing fuel flow to get higher cylinder temperatures. The PA-2 was introduced at about the time EMD began offering the twin-engine E-8 at the same power level. By this time most lines had settled on a distribution of power from various builders. Some had little or no Alco power, but others maintained an interesting mix. Most PA/PB orders went to former customers who liked their PA-1s.

In a repeat of the Freedom Train Operation, the first two PA-2s were assigned to a 15-month nationwide tour by General Electric entitled "More Power to America." The locomotives were painted orange-red with the two corporate logos on the nose and the train's name on their flanks. After this service ended, the units were sold to NYC in 1951.

Figure 6.5. Two views of the famous PA model in action. GM&O's PA-1 No. 292 first gained fame as No. 1776, leading the *American Freedom Train* across the nation between September 1947 and January 1949. A memorial plaque of the honor is visible above the front truck in this early morning view of the *Gulf Coast Rebel* leaving Meridian, Mississippi, in June 1954. Below, Missouri Pacific No. 72 leads the northbound *Texas Eagle* downgrade toward a stop at Austin, Texas, in May 1963, less than a year before retirement. J. Parker Lamb (both).

Although a distant second in overall production, Alco did find a few devoted users of its postwar passenger power. The top two were Missouri Pacific's 36 PA/PB units, which were equal to its roster of EMD E-models, while the grand champion was Southern Pacific, which amassed an Alco fleet of 66, over twice the number of EMD machines. Moreover, it was the only road to completely upgrade all of its PA-1s with the 244D engine in the early 1950s, and six years later it launched a complete overhaul of the electrical systems in all its PA/PBs. This included installing modular electrical cabinets with plug-in components and the installation of conduit troughs that consolidated and simplified wiring. The final tally for PA/PB-2 production was 90 units (only 8 Bs), of which 3 were sent to a Brazilian railway.

But the final chapter of the PA design and its revered aesthetics is still being written. As mentioned earlier, two PAs were returned to the United States in 2002 after being purchased from the National Railways of Mexico. They were in a group of A-B sets delivered to the Santa Fe as part of its second order in 1947. After their two decades of ATSF service ended, a pair of A-B sets were purchased by the Delaware & Hudson and refurbished for passenger service. This move was prompted by D&H president Carl Sterzing Jr.'s longtime affection for these classic machines, along with two Baldwin RF-16 Sharknose units he saved from a scrap yard. The PAs, numbered 16–19, were painted in a blue bonnet scheme in which the red Santa Fe style was reconstructed in D&H's standard light blue with gray and yellow trim. In a 1975 modernization move, the original model 244 power plants were replaced by more powerful 251 engines.

After a change in D&H's management, the Alcos migrated south of the border in 1978 for another long stint of service, but they were ultimately put in a storage track and forgotten. In the 1990s, after word drifted back that two PAs were still in salvageable condition, a lengthy period of negotiation ensued. This finally ended with their purchase by an American consortium, led by a well-known historic locomotive rebuilder. The lengthy restoration process was in progress in 2005. One of the units (originally ASTF 62L) will be restored to operating condition as Nickel Plate No. 190. The other's final destination has yet to be decided.

Figure 6.6. The only remaining examples of the original PA car bodies are two of the four ex–Santa Fe units bought by Delaware & Hudson and later reengined before being sold to the Mexican rail system. While still in D&H's blue and silver, PA No. 17 and its three mates leave Whitehall yard after an excursion trip in May 1977. Jim Shaughnessy.

Figure 7.1. The F3 model was EMD's first freight unit after World War II. Monon's No. 84A (1947), heading out of Lafayette, Indiana, in 1958 with *Thoroughbred*, was typical of this popular model, many of which were equipped with steam generators for passenger operations. High fan-housings on the roof were characteristic of units constructed before 1948. J. Parker Lamb.

CHAPTER 7

POSTWAR SHAKEOUT

With its momentum from prewar developments of passenger units and wartime success with the pioneering FT design, Electro-Motive began its next chapter as soon as WPB restrictions were lifted in 1945. Indeed, the FT's successor, the F3 model, began production in July 1945, before the official end of World War II.

Carrying virtually the same bulldog-nosed car body as the FT, the new unit was powered by the upgraded 567B engine (1,500 hp). The earliest F3 A-units displayed three side portholes rather than the FT's four, whereas later versions replaced the middle window with small horizontal vents, and some even included an overlay of heavy wire mesh in this area.

With its plants going at full blast to meet pent-up demand, immediate postwar production at LaGrange between 1945 and 1949 totaled 1,887 F3 A and B units (including a small group of F2s produced for four months in 1946). During the same period, its E7 passenger unit production added an additional 510 A and B units. By 1949 production began on the F7 and E8 models, which would also be turned out in prodigious quantities until 1953, with total counts of 3,950 for F7s and FP7s and 457 for E8s (both A and B). The FP series, slightly longer to accommodate a large steam generator, recognized the inherent versatility of a diesel unit as both a freight and passenger locomotive.

The final versions of EMD cab units (F9 and E9) were introduced in 1954 using the 567C engine (1,750 hp). The passenger model was built until 1963, while cab freight models would be terminated in 1957. Production totals for the (-9) models were 175 for freight and 144 for passenger units. Clearly, Electro-Motive's total output for the first postwar decade represented a new level of dominance for the GM subsidiary.

The above scenario implies that all was not well with some of the remaining diesel locomotive builders, who would be the financial victims of the postwar shakeout.

When it came to the heavy machinery firms, Baldwin, Fairbanks-Morse, and Lima, it usually was a case of their locomotives being "too big, too different, or too late." Even though two of these companies had been among the nation's leading steam power producers, none was able to challenge Alco for the second rank in production.

While all of them hired competent engineering designers and executives, each had its own traditions and personalities. In their short tenures as diesel builders, these underlying traits dictated their different approaches to new product development and subsequent manufacturing processes. Even though their locomotives were not without strong supporters, they were not able to increase market share due to the intense competitive

Figure 7.2. SAL No. 4028 (1948) illustrates the F3 model with wire mesh covering the side windows. Produced for a year starting in mid-1947, these were commonly referred to as "chicken wire" units. Shown at Tampa in 1950. J. G. Lachaussee Collection.

Figure 7.3. Following its record-setting purchases of FT units, Santa Fe continued its rapid dieselization by acquiring F3s and F7s in large numbers. In front of the massive shops in Cleburne, Texas, a late afternoon sun highlights the blue-and-yellow F7 No. 204C (1949) and its passenger-hauling cousin, the red-and-silver F7 No. 342 (1953). J. Parker Lamb.

Figure 7.4. Great Northern E7 No. 512 (August 1947) was one of the road's last E-units and, like many other GN diesels, was fitted with a snowplow pilot for winter navigation. Shown at St. Paul in July 1968. J. Parker Lamb Collection.

Figure 7.5. On the Southern Railway, local passenger runs were entrusted to FP7s such as No. 6147 (November 1950) while streamliners, in this case the *Southerner*, were led by lean racers like E8 No. 2924 (September 1951). These two spotless units pause briefly in Meridian, Mississippi, in 1953. The FP7 unit would last into the Norfolk Southern era (1988). J. Parker Lamb.

Environment. However, as will be seen, they were eventually doomed by events far from the design offices and erecting bays.

Baldwin Locomotive Works

Although their VO-series engines worked well in early switcher units, it was clear that the postwar market for road locomotives would require a more rugged design. The new 600-series inline engines represented further refinement of the VOs with a strengthened frame, new cylinder heads, and improved valves and pistons. Three variations were produced starting in late 1945: the 660-hp Model 606-NA (six-cyl. normally aspirated), the 1,000-hp Model 608-NA (eight-cyl. normally aspirated), and the 1,500-hp Model 608-SC (eight-cyl. turbocharged). Elliot superchargers were used and all engines were rated at 625 rpm. In parallel with Alco's experience, there would be later teething problems with the supercharged engines. While this engine redesign program was in progress, the company was also planning a new breed of roadworthy locomotives.

During the last phase of steam locomotive development, BLW was fascinated with extra large locomotives, such as the series of Pennsy duplex drive engines (Q1, Q2, T1, S1). Although they represented a new concept in steam locomotive design, their overall performance was satisfactory for such semi-experimental machines. Indeed, their major problem was based not on technology but on their accommodation to contemporary operating practices. In essence, they were too big for their time, requiring special handling by both shop and operational people.

EMC's 1939 development of its E-series passenger locomotives occurred during the period when there was considerable concern about union demands that each unit include a two-man crew. As was noted, this led to some unusual numbering schemes. But for Baldwin, with its penchant for huge machines, the (not unexpected) reaction was "We will build a bigger diesel that provides sufficient power in one unit." Unfortunately, this stance overlooked one of diesel-electric technology's greatest assets, the operating flexibility of relatively small power modules (units). The company's initial planning was driven by a concept credited to Baldwin's Swiss-born engineer Max Issl, who envisioned another type of modular locomotive. In this case a series of small engine-generator sets would be mounted crosswise to the car body and would drive only one axle. Each locomotive model would enclose a different number of these modules and thus produce the desired flexibility in size. Issl's work was represented on three patent applications (1939, 1942, 1943) for large passenger diesels having at least 4,000 hp. Construction of the first unit (No. 6000) began in January 1941 and, from our present perspective, can only be characterized as a supersized Rube Goldberg invention.

Supported by a 2–D+D-2 running gear characteristic of electric locomotives, plans called for using *eight* 750-hp V-8 engines (Model 408), thus providing a whopping 6,000 hp. The massive body sported a long, slightly slanted nose, but its most distinguishing appearance feature was a pair of porthole windows above each engine. The ensemble of 16 round windows reminded observers of an ocean liner on wheels. Needless to say, completion of this huge machine during the period of WPB restrictions was difficult, and after a few months of unimpressive road tests, No. 6000 was stored in November 1943. Surprisingly, the running gear from No. 6000 was resurrected in a new, more modestly designed model known as DR-12-8-1500/2 (for Diesel, Road, 12 axles, 8 driving axles, two 1,500-hp engines). Its 24 wheels immediately produced the nickname "centipede," and that's what it was called until the end. With 45,300 pounds of tractive effort, the centipede was the most powerful diesel up to that date.

Powered by a pair of Model 608-SC engines, the 1945 demonstrator, also numbered 6000, barnstormed over eastern lines before being purchased by the Seaboard Air Line in December 1945. Assigned initially to passenger service in Florida, it was eventually trans-

Figure 7.6. The Pennsylvania Railroad purchased the largest fleet (24 units) of Baldwin's twin-engine 3,000-hp centipede model of 1947–48, whose 24 wheels occupied the entire underframe. A pair of these drawbar-connected behemoths, led by No. 5832, is shown at Coshocton, Ohio, in December 1948. Louis A. Marre Collection.

ferred to freight work due to its lugging ability. Soon the Mexican National Railways (NdeM) ordered 14 units, Pennsy bought 24, and SAL returned for 13 more. After these were delivered in July 1948, no further orders were received, and Baldwin's Big Diesel period was finished. Comments from users of the centipede were remarkably similar to those regarding Pennsy's Baldwin-built duplex steam power: "Too big and too unreliable."

However, Baldwin had not put all of its efforts into the oversized centipede design. A parallel development program also produced a more conventional 2,000-hp passenger unit, running on A1A trucks. Designated as DR 6-4-2000, they were powered by two 608-NA engines and Westinghouse electrical gear. The first two orders were placed in November 1946 (Central of New Jersey) and December 1946 (GM&O). The latter units represented the convention layout, with both generators facing the cab, whereas the CNJ locomotives, designated for commuter trains, carried double-cabs to eliminate turning. This necessitated a new internal arrangement in which the generators faced each other and led to a revised model designation as DRX 6-4-2000.

Although the frontal styling of these two-trucked passenger units were similar to the centipedes, they quickly received another nickname from diesel power chroniclers. The name was based on the unusual cab location, displaced well below the maximum height of the car body. Observers described this as akin to a baby's face on which features are small in relation to the head. Thus the locomotives were denoted as babyface cab units. Early versions included a transition cowling between the central car body and the cab area, while later units featured a sharp break between the two heights.

By 1947 another round of engine upgrades found the six-cylinder output increased from 660 to 1,000 hp with an Elliot supercharger. The new 606-SC thus could replace the more expensive 608-NA. Used mostly for switchers, the first mainline application for this engine was a special Chicago & North Western version of the dual engine RD 6-4-2000 for use on branch line trains. The rear engine room in this DR 6-2-1000 was converted to a baggage compartment. The 6-2 designation reflects the use of an A1A-3 wheel arrangement (two powered axles in front). Similar configurations for low-density service had been produced in 1940 by EMD for Rock Island and Missouri Pacific. Not surprisingly the largest order for

Figure 7.7. In top photo, both of GM&O's Baldwin 6-4-20 units (Nos. 280–81) lead the *Gulf Coast Rebel* as it leaves Meridian, Mississippi, on the last leg of its St. Louis–Mobile run in 1952. New York Central's fleet of Baldwin DR 6-4-1500 units was eventually given EMD engines and MU-ed with F3s on Big Four lines' passenger trains between Cincinnati and Cleveland. No. 3506 is speeding northward after leaving Dayton in May 1956. J. Parker Lamb.

Figure 7.8. The Sharknose styling of Pennsy's Baldwin passenger units was carried over to its freight cab units, illustrated at the top by a 1958 view of a coal train near Cleveland led by RF-16s, Nos. 2005 and 9743 (1951). At the bottom, a southbound NYC freight, led by RF-16 Sharknose unit No. 3815 (1952), rumbles by the Maitland tower near Springfield, Ohio, in November 1960. Peter Eiesle (top), J. Parker Lamb Collection (both).

the conventional 2,000-hp model (nine A-B-A sets) was from Baldwin's biggest customer, the Pennsy, which decided to bring in its favorite industrial stylist, Raymond Loewy, to give the unit a new look, one that would stand apart from the competition. To say that he succeeded would be a gross understatement, as his design used a forward slanting nose with sculpted sides similar to the T-1 duplex steam locomotives. Not surprising was the immediate nickname, Sharknose, for these striking units. Of the total production (36) of 2,000-hp passenger units, PRR's 27 were by far the most successful in mainline performance. Thirty-six of these 2,000-hp units were sold to four U.S. roads and three to the Mexican national railroad.

In a last-ditch effort to compete with EMD's F-3 in passenger service, BLW offered its DR-6-4-1500 in 1946. Only 53 feet in length, some 20 feet shorter than the twin-engined GM&O units, these babyface units were powered by a single 608-SC engine, and they rode on Commonwealth A1A trucks. Unfortunately for Baldwin, only two roads opted to purchase this model. New York Central bought two A-B-A sets in February 1946, and SAL ordered three A-units seven months later. The SAL machines spent 17 years operating on secondary lines in Florida. NYC's fleet was soon equipped with high-speed trucks and eventually redesigned at EMD before being retired in 1958.

While Alco was gearing up to produce its Black Maria freight units, BLW was designing its own B-B, the DR-4-4-1500, a babyface car body powered by the 608-SC engine and riding on General Steel Castings Type B trucks. Although Central Railroad of New Jersey ordered five A-B-A sets in September 1945, production problems delayed the first deliveries until November 1947. By then the New York Central had ordered eight As and four Bs and Missouri Pacific had ordered four A-B-A sets, all for delivery in 1948. The next version of this freight unit, while offering only minor improvements to the internal machinery, was delivered in the Pennsy-inspired Sharknose car body. Not surprisingly, PRR's order for 34 A-B sets for delivery in 1949–50 was the major one for this configuration, with only Elgin Joliet & Eastern receiving two A-B sets in 1949. In all, the DR-4-4-1500 deliveries spanned 32 months and totaled a disappointing 105 units bought by five roads.

However, the Sharknose configuration was extended to the next model, newly designated as the RF-16 (for Road Freight, 1,600 hp). Powered by a new 608A engine, successor to the 608-SC, it was expected to compete with Alco's FA-2 (1,600 hp) and EMD's F-7 (1,500 hp). The first RF-16 buy was in July 1950 for 14 As to B&O. Pennsy was next with 44 A-units and 16 B-units in August, and in March 1951 NYC ordered 18 As and 8 Bs. Good performance by these units ignited a round of reorders, so that total production was 109 As and 51 Bs for 1951–53. But, more significantly, the RF-16 brought down the curtain on Baldwin diesels with streamlined car bodies. American railroads had discovered a more utilitarian configuration for its diesel power (see chapter 8).

Fairbanks-Morse

Begun in 1830 as a maker of scales, the E. & T. Fairbanks Company of St. Johnsbury, Vermont, was taken over in the 1870s by Charles H. Morse after his reorganization of a new headquarters office in Chicago during that city's post-fire reconstruction. Eventually the company expanded into tools and food processing, and in 1885 it acquired a windmill company, Eclipse Wind Energy Company, as well as the Beloit Wagon Works. It was the windmill business that led the company into pumps and other reciprocating machines, including stationary steam engines. Purchase of the George Sheffield Company expanded F-M into the railway supply field while, in 1893, Morse hired the patent holder for the first two-cycle gasoline engine and began a production program. The 1906 acquisition of the Three Rivers Electric Company of Michigan brought electric motors and generations under company's broad umbrella.

In 1922 F-M hired F. P. Grutzner, a onetime apprentice to Rudolf Diesel, who later migrated to the De La Vergne engine company. He and fellow German engineer Heinrich Schneider, working under chief engineer Larry B. Jackson (later of Alco), pioneered the early success in developing the opposed piston concept, developed in response to a 1932 U.S. navy request for a submarine diesel. The OP design, in which each cylinder contains two pistons that turn crankshafts at either end, thus provided almost twice the power per unit of engine volume. This configuration (low-volume, high-power density) was an ideal match for U.S. submarine propulsion.

While the navy was delighted with OP engines, the initial railroad applications to passenger-carrying motor cars were not successful. This early engine was a 5 × 6 (bore, stroke) machine that produced too many maintenance problems and thus soured many in the rail industry on the OP engine. But this early misadventure taught the company some good lessons, and its second railroad engine, an 800-hp, five-cylinder (ten pistons) 8 × 10 machine, was good enough to power six Southern Railway shovelnose motor cars (rear A1A truck had only power) built by the St. Louis Car Company in 1939. These were the first commercially successful OP engines for rail use, and they served satisfactorily on the Southern cars until they were retired in 1955.

The company's long-range plans for a line of locomotives had originated in 1935 with the hiring of John K. Stotz, an electrical engineer from Westinghouse who specialized in traction motor design. By 1940 preliminary designs were prepared for a 1,000-hp switcher and a 2,000-hp dual-purpose cab unit whose car bodies would be constructed at GE-Erie. But the subsequent WPB restrictions denied F-M any railroad production during the war because its submarine engine output was much more critical to national needs than a rail market being adequately served by three builders (Alco, Baldwin, and EMD).

Despite the big production lead by existing locomotive companies, F-M was banking on the strong performance of its revolutionary (for rail service) OP engine when it obtained WPB approval to start railroad production during the summer of 1943, soon hiring John W. Barriger as the manager of its Diesel Locomotive Division. As its 1,000-hp switcher began fabrication, the nearby Milwaukee Road quickly signed up as the first customer. The F-M engineering staff was led by Grutzner (while Stotz was on army duty) and included a former professor, L. E. Endsley. Since time was their greatest enemy (due to wartime delays), the company decided to purchase from Baldwin five sets of running gear used for its switchers (underframe, trucks, and traction motors). They then mounted their OP engine (800 rpm, 8 × 10, six cylinders) and a Westinghouse-built compatible generator under their own car body, based on a Loewy design, and soon the first Fairbanks-Morse locomotive was ready in August 1944, only a year after the company received clearance from the WPB. Due to the extreme height of the twin-crankshaft engine, the hood was as high as the top of the cab. This car body would be the standard for F-M switchers until 1952. While the first and last of the five Baldwin-based units went to Milwaukee Road, the intermediate ones were bought by C&NW, Santa Fe, and Union Pacific. By August 1945 the company was fabricating its own main frames to the same general specs as the cast frames from Eddystone. After producing 195 of these H-10-44 models between 1944 and mid-1950, the 1,200-hp H-12-44 model (with rpm increased to 850) was produced until 1961 with 306 U.S. units and 30 Canadian units sold.

Once switcher production was moving, the primary focus of designers was a large dual-purpose cab-type unit riding on A1A trucks. Due to Endsley's attempt to operate the ten-cylinder, 2,000-hp engine at nearly the same temperature as a submarine unit (surrounded by cold water), the extra cooling capacity required a unit nearly 65 feet in length, only 6 feet shorter than EMD's E-7 model. With limited plant capacity, F-M turned to GE-Erie to fabricate the car bodies, which were based on another Raymond Loewy design with his usual long nose (originally 15 feet long but later reduced to 9). The first A-B-A set for Union Pacific was rolled out of Erie in December 1945, at about the same time as Baldwin's centipede models and six months ahead of Alco's PA-1. Soon the Santa Fe ordered an A-B-A set, and Milwaukee Road bought six A-B-A sets to cover its new *Olympian Hiawatha*

Figure 7.9. Fairbanks-Morse's first switcher model, the H10-44 of 1944, featured a Raymond Lowey–styled roof overhang in the rear that was reminiscent of early Camelback steam locomotives. At top, Milwaukee Road No. 755 is at Green Bay, Wisconsin, in 1970. In contrast, Baltimore & Ohio No. 9726 (1957) was a late version of the H 12-44, which carried a shortened car body as a cost reduction measure. Shown at Riverside yard in Baltimore in August 1972. J. Parker Lamb Collection (top), James Mischke Collection.

train. The first group of units bought for freight was KCS, which initially tried them in 8,000-hp lashups. However, their enormous power, aggravated by slack action on long trains, led to many broken couplers on the road's hogback profiles in Arkansas and Kansas. Thus KCS settled for three A-B-A sets. Pennsy's order of 48 units (36 As and 12 Bs) was the largest group of the eventual total of 111 Erie-built units constructed before discontinuance in February 1949.

Unfortunately, the compact and efficient opposed piston engine began to experience serious problems during its first five years of service, primarily due to piston and crankshaft

Figure 7.10. To provide more power on its hilly routes in Missouri and Arkansas, Kansas City Southern bought three A-B-A sets of 2,000-hp A1A cab units, powered by Fairbanks-Morse engines but built at GE's plant in Erie. At top, the 60-set is seen at Pittsburg, Kansas, in 1950. The smaller units of F-M's C-Liner series represented its last products before the road-switcher era. The 1,600-hp CFA-16-4, illustrated by Milwaukee Road No. 26C (1952), was at Bensenville, Illinois, in April 1967. Louis A. Marre Collection (top), J. Parker Lamb Collection.

failures from the higher stresses developed in rigorous railroad service as compared with the relatively benign operation at constant speed in submarine applications. As we have seen with other such cases, it was difficult for many engineers to understand that one could not take a well-performing diesel engine from marine or stationary usage and transfer it without change to the much rougher railroad environment. Thus a fresh batch of new engineering talent arrived from competitive companies in 1947 to take charge of improving Fairbanks-Morse's engine performance and reputation.

The incoming division manager was V. H. Paterson of Baldwin, replacing Barriger, who left to be president of the Monon, while from EMD came John Wiffenbach, engineering manager, and three other engine and electrical experts. Their first tasks were to plan a new line of road power and the plant in which to assemble it, since the GE contract would expire in 1949. Although the new plant was completed at Beloit in 1948, much of the design and engineering was done in the Chicago offices. The new engineering group decided that F-M should produce a line of dual-purpose cab models with improved engines. By switching to aluminum main bearings on the crankshaft as well as improving its counterbalancing, the engineering staff anticipated that engine failures would become past history.

The new designs, named by Chicago engineer Henry Schmidt, were called the Consolidation Line (later shortened to C-line). They would share a 56-foot car body based on the styling of the much longer Erie-built units. The concept was to offer a range of power from 1,600 to 2,000 to 2,400 hp using OP engines of various sizes (8, 12, and 16 cylinders) as well as a new truck design. Freight models would have B-B running gear while passenger units would use an unusual B-A1A arrangement. The first C-liners were produced in December 1949, but their fate parallels that of Baldwin's RF-16 Sharknoses in that, by the time they were ready for marketing, the peak demand for cab units had passed. Thus by 1955 only 195 units had been constructed for seven carriers, including 99 units for the U.S. roads and 66 for the two Canadian roads.

Lima-Hamilton

Of the large commercial steam builders, Lima Locomotive Works' plant was the last one standing, producing its last ten locomotives (Nickel Plate Berkshires) in March through May 1949. Lima also contracted with Alco, already converting to diesel work, for construction of tenders for that company's final seven NYC (P&LE) 2-8-4s. As surprising as it may seem, on the same day (May 13) that the last Nickel Plate steam engine was completed, a black 1,000-hp diesel switcher (numbered 1000) also emerged from the Lima plant. While it still carried the famous red diamond builder's plate, the company name emblazoned on the hood was not Lima Locomotive Works but Lima Hamilton Corporation, reflecting a merger with General Machinery Corporation almost two years earlier (October 1947).

General Machinery was itself a product of the consolidation of numerous companies in December 1928. One of General's major components was its Hamilton engine division, which like F-M was a major builder of navy diesels (not using the OP design) during World War II. On the surface it seemed like an ideal merger, with Lima having extensive fabrication facilities and a high standing in the railroad industry whereas Hamilton was seeking a wider market for its diesels in this sector. But when the decision was made to produce diesel locomotives rather than steam, there were many diehard steam proponents that decided to leave Lima.

The L-H locomotive program was under the direction of engine expert F. J. Geittman (formerly of Alco) with car body design led by M. J. Donovan (Lima). The power plant was a modification of Hamilton's successful marine engine, an eight-cylinder, turbocharged inline machine with 8.75 × 12 cylinder dimensions. The railroad version had a 9-inch bore

Figure 7.11. Lima-Hamilton's largest locomotives were 22 twin-engine, 2,500-hp transfer units built in 1950 for the diesel-hungry Pennsylvania Railroad, whose No. 5680 was at Columbus, Ohio, in June 1956. J. Parker Lamb.

and was beefed up structurally (crankshaft, piston rods, etc.) to better withstand harsh railroad conditions. In many respects the first diesel locomotive design displayed the same attention to excellence in details as was characteristic of Lima's steam locomotives. These features included extra cooling for internal components (such as manifolds), interchangeable exhaust and intake valves, heat-treated pistons, a counterweighted crankshaft, and internal cooling passages in connecting rods, pistons, and wrist pins. Many of these design changes were in response to early engine problems of the other builders.

The car body and trucks were essentially the same as used by other manufacturers (especially Alco), except for the large, canted number boards at the front (similar to those NKP Berkshires) and the four side windows on the cab. Successful demonstration tours suggested that additional visibility was needed on the rear wall of the cab, but the general consensus was that No. 1000 was a strong puller. Unfortunately, Lima-Hamilton could not overcome its late entry in to the rail field and was able to produce locomotives for only two years, before itself being consumed in 1947 by the same merger movement that created it. During its short production life, L-H turned out four sizes of switchers: 750 hp (6 units), 800 hp (23), 1,000 hp (38), and 1,200 hp (69), for a total of 136 units. In addition it sold only a few samples of its planned line of larger power, including a 1,200-hp light road switcher (16 to NYC) along with 22 of its largest model, a twin-engine, 2,500-hp transfer locomotive for Pennsy.

With many L-H customers satisfied with the performance of their machines, there were enough orders on the books in 1951 for the Ohio company to pull even with or even surpass, Baldwin as the fourth producer. But deep behind the scenes, Baldwin's owner Westinghouse was in secret negotiations with Lima-Hamilton regarding a merger that was consummated in October 1950. As might be expected, L-H's designs were discontinued and all work focused on Baldwin locomotives, now marketed under the name Baldwin-Lima-Hamilton. Although the Lima name did not disappear, its eight decades of meticulous craftsmanship were gone forever.

Figure 8.1. Although a 1947 vintage F3 model was leading this Central of Georgia train at Leeds, Alabama, in August 1957, such cab units were being replaced rapidly by the more utilitarian and versatile hood-type diesels (GP9s and RS-3s). J. Parker Lamb.

CHAPTER 8

ROAD SWITCHERS TAKE OVER

After diesel locomotives became the primary workhorses of the industry, it was inevitable that many of its early appearance features would be subject to scrutiny, with a view toward simpler designs and less costly fabrication. Especially expensive were the streamlined car bodies that required the piecing together of sheet metal panels that were stamped with compound curvature. Somewhat surprisingly, industry pacesetter EMD would not be the early leader in developing a more utilitarian car body. Indeed, it was the last to introduce a viable competitor into the four-builder race. Previous mention has been made of Alco's pioneering road switcher model (RS-1) introduced during World War II and its immediate success in foreign and domestic service. This configuration, later denoted as a hood unit in contrast to a cowl or cab unit, would become almost universal on American railroads. But it would take over five years beyond the war for these designs to be fully developed and accepted throughout the industry.

American Locomotive

With its extra years of wartime experience, Alco sprinted from the gate with its RS-2 model, styled by Ray Patten in parallel with his work on the more famous PA and FA models in late 1945. Taking the basic utilitarian configuration of the RS-1, he rounded all the corners (cab and hood) to present a semi-streamlined appearance, quite appropriate for a locomotive whose emphasis was on working the main lines rather than displaying its sleek lines.

From the beginning Alco had planned to offer the 1,500-hp RS-2 (244 engine) on A1A, B, or C trucks. Surprisingly, the first order from the Milwaukee Road in August 1945 was not for a typical B-B unit. Rather, the road wanted 18 RSC-2 models with A1A trucks for branch line service in Wisconsin and Michigan. The first unit was delivered in October 1946 for test runs, with the remainder following into January 1947. The first RS-2 (B-truck) was delivered to the Detroit & Mackinac in November 1946, soon after the initial RSC-2.

Most early customers of road switchers were smaller lines such as Toledo Peoria & Western, Alton & Southern, Lehigh & New England, Union Railroad, Elgin Joliet & Eastern, and Belt Railway of Chicago. Among larger lines buying blocks of RS-2s were Southern (30), Delaware & Hudson (27), NYC (23), Erie (21), NH, and B&M. The latter four roads used these new units to convert their commuter services. The D&H, which

Figure 8.2. Three-axle trucks allowed heavy diesel units to use lines with light rail, as depicted by this scene of an eastbound Seaboard train leaving Montgomery, Alabama, in July 1954 behind two RSC-2s leading an RSC-2M with modified truck frames that eliminated the idler axle. Note that these early units were still riding on friction bearings. J. Parker Lamb.

served the Alco plant and was a longtime customer, completely dieselized its mainline and local trains with RS-2s, and Seaboard Air Line, with hundreds of miles of light rail branches in the Deep South, became the largest buyer of the RSC-2 (36 units) plus an additional 24 RS-2s. Total production by the end of 1949 was RS-2 (315) and RSC-2 (76).

With competitive pressures mounting to upgrade its road switcher line in the mid-1950s, Alco introduced the 1,600-hp RS/RSC-3 models. These used the upgraded 244D engine and a larger fuel tank relocated to a position between the trucks (under the main frame). Improved truck designs also came from Adirondack and GSC, along with a provision for dynamic braking in the short hood. Although virtually identical in general appearance to the RS-2 line, there were minor changes in details, such as battery box covers and hood side louvers/filters. Between 1950 and 1956, a total of 1,265 RS-3s were produced for U.S. lines plus another 98 for Canada and 7 for Mexico. They were especially popular for commuter service in Chicago (Rock Island) and the New York City–Boston area (B&M, CNJ, Erie, Long Island, and Reading). In contrast, the A1A version (RSC-3) sold only 11 U.S. and 8 Canadian units.

To counter this trend, the company introduced the C-C version (RSD model), which was offered between 1952 and 1956. The RSD-4 was marketed for only two years (1951–52), and only 36 units were sold, while the RSD-5 fared much better with 167 U.S. and 37 Mexican units. It is important to remember that, by this time, all of the RS-3 and RSD models were in direct competition with EMD's newly developed road switcher models. In 1954 Alco produced a high-horsepower C-C model in response to a similar unit constructed by Fairbanks-Morse. Known as the DL600 (RSD-7), it used a beefed-up 244G engine (2,250 hp) and featured the highest capacity dynamic brake of any freight locomotive then available. Its hoods featured notches at the corners, constructed so that the number boards were at the proper angle with sand box covers directly below. This notched nose was both functionally efficient and visually attractive, and it became an Alco standard.

Two demonstrators (numbered DL 600 and 601) set out in early 1954 on a road tour that produced only three buyers for the production model DL 600A. For this model the engine

Figure 8.3. A Southern Pacific RSD-5 (April 1955) carries a four-digit Pacific Lines number, but was rambling along with a local northwest of Austin, Texas, in July 1985. A number of roads set up their Alco road switchers to operate normally with the short hood forward. J. Parker Lamb.

was further improved to produce 2,400 hp, and an aftercooler was added to reduce air temperature leaving the turbocharger and thus increase engine efficiency. The buyers of this model were C&O (12), Santa Fe (10) and Pennsy (5). In addition, ATSF bought the two DL 600 demo units. Although not a big seller, the RSD-7, which represented the final usage of the Model 244 engine, set the stage for Alco's last generation of locomotives, powered by a new engine.

Baldwin Locomotive Works

It is somewhat surprising that the record for delivering the first 1,500-hp hood unit goes to the company from Eddystone, whose road switcher program began in the summer of 1945 when it was contacted by the master mechanic of a Mississippi shortline, the 168-mile Columbus & Greenville. He requested bids from Alco, Baldwin, and EMD for a general-purpose 1,500-hp locomotive with A1A trucks for its light roadbed and wooden trestles. Baldwin was the only builder to respond seriously, as its engineering staff had already been planning to use the 608-SC engine as the basis for a heavy road switcher. An order for five units was placed in September 1945, and the first unit (model DRS 6-4-1500) was delivered a year later.

Thus the nation's first 1,500-hp road switcher began service on October 8, 1946, only six days after Alco's first RSC-2 had been completed but not yet put into service by the Milwaukee Road. Its squared-off engine hood, almost cab-high, gave the Baldwin unit a heavy look (in comparison with the RSC-2) that was accentuated by the cast side frames of the GSC Commonwealth trucks. Although there were the usual teething problems with the C&G unit, the road later received four more to complete its order by January 1947 and even added a sixth Baldwin road switcher (AS 416) in 1951. It ran only Baldwin power on its

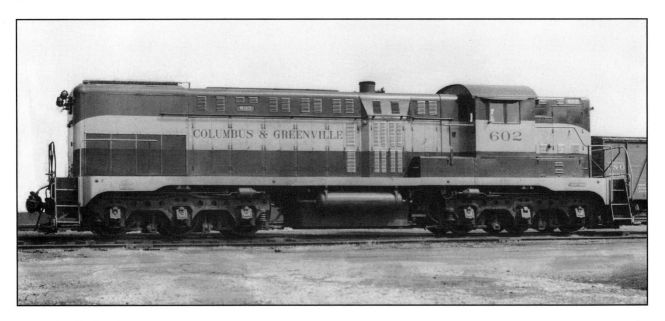

Figure 8.4. In October 1946, Mississippi's Columbus & Greenville put into service the nation's first 1,500-hp road switcher, Baldwin model RS-6-4-1500, numbered 601. Delivered two months later was No. 602, displaying its bright orange and green paint at the Columbus shops in December 1954. J. Parker Lamb.

main line until newer units were purchased in 1965. Indeed, two of the original six units remained active until the early 1980s.

While assembling the C&G units, BLW received a 52-unit order for specially equipped 1,500-hp road switchers for use in French North Africa on a line through the Sahara Desert. Such service, of course, required additional air-cleaning equipment and sealed compartments in order to minimize any sand ingestion by the machinery. Before starting to work on this order, Baldwin constructed an official demonstrator unit, painted dark red with silver striping and carrying the number 1500. Starting in November 1946, it was tested in both freight and passenger service on New Haven and Central of New Jersey.

The first order for Baldwin's 1,500-hp B-B trucked hood unit (equivalent to Alco's RS-2) came from Western Maryland in July 1946. Following this, the Soo Line ordered eight of these DRS 4-4-1500 units, and the NYC acquired two for testing on secondary passenger trains. Two more regional lines from the South provided the next big orders. In January 1947, the 339-mile Norfolk Southern Railway, a satisfied user of Baldwin switchers, dieselized its main lines with ten DRS 6-4-1500 models. Four months later, the 144-mile Savannah & Atlanta bought eight of these light rail versions for its mainline trains. The heaviest 1,500-hp models (C-C trucks) debuted in February 18, 1948, when three DRS 6-6-1500s were delivered to the Chicago & North Western. It would be three years before Alco would market a comparable unit. In the meantime Baldwin sold 83 of these six-motor units to 13 lines, including 16 to three steel company roads and 7 to three regionals or shortlines. Baldwin's early successes in the road switcher arena (147 U.S. locomotives plus 61 foreign sales) suggested that it might be able to become a stronger competitive.

Eddystone followed Alco's trail in the immediate postwar period by marketing a 1,000-hp road switcher based on a 660-hp export model that sold 106 units (1946–48). The DRS 4-4-1000 (B-B) used the 606-SC engine and sold 22 units in three years (1948–50). The improved 1,200-hp RS-12 model (608A engine) was more popular, with 50 sales between 1951 and 1954. In 1950 the company decided to discontinue its pioneering DRS models and introduce an improved line that included dynamic brakes, roller bearings, dual control stands, new brake controls, and a water-cooled air compressor. Designations for B, C, and A1A trucked versions became AS-16, AS-416, and AS-616, although their external appearances were virtually unchanged from the previous models. With 83 orders in 10 months for the AS-616, the general popularity of these models reflected a nationwide need for new power, especially road switchers (a total of 4,473 orders were placed in 1950).

Figure 8.5. The original Norfolk Southern Railway operated only Baldwin power on its primary lines. A Charlotte-bound train leaves Raleigh, North Carolina, in June 1962 behind a quartette of AS416s, the second of which is a high-hood version from the last batch built in 1955. J. Parker Lamb.

Orders for the AS series grew rapidly in 1950, reaching a peak in 1952 and 1953, but then a slow decline began. The last of 222 AS-616s was produced in January 1955, AS-16 ended production at 127 units six months later, and the final models of the AS-416 were completed in December 1955 after a total output of 25 units. Among the top buyers of Baldwin road switchers were Southern Pacific (96), Reading (43), C&O (42), Erie (34), and the Norfolk Southern Railway (27). One of the major influences on Baldwin's production was Westinghouse's 1953 announcement that it anticipated an end to production of railroad traction equipment. Although Baldwin stocked up on WEMCo parts for future production, it later had to modify its designs to accommodate GE equipment.

The future became clear in 1954 when Baldwin-Lima-Hamilton chairman George Rentscher declared the company's intention to reacquire all B-L-H shares held by Westinghouse. He also noted that 80 percent of the company's business was now heavy machinery with railroad orders accounting for less than 20 percent. The official end to Baldwin and Lima as locomotive builders came in 1956 after a total of 3,208 diesel units had been constructed at Eddystone.

Figure 8.6. Baldwin
locomotives exhibited
good performance at low
speed, and thus many were
employed as hump pushers.
At top, a 1962 view of SAL's
Hamlet yard finds an S-2
teamed with an RS-12, both
built in 1952. Southern
Pacific's Englewood hump
in Houston soared to a
height of 50 feet in order
to clear a main line.
Thus three-unit pushers
were usually needed, as
illustrated by this July
1964 view of an A-B-A
combo of Baldwin
DRS-6-6-1500 units, all
equipped with dynamic
brakes. J. Parker Lamb.

Fairbanks-Morse

This ambitious group was the third builder to recognize the hood unit's strong potential, displaying two new B-B models at the 1947 Atlantic City Show. The more traditional unit was a 1,500-hp model (H-15-44) with the usual short hood, but the head turner was a 2,000-hp red-orange locomotive (H-20-44) that, with its compact ten-cylinder O-P engine, resembled a large end-cab switcher. Named Heavy Duty, the large unit impressed many rail observers with the enormous power packed into such a small space. Indeed, Union Pacific literally bought the demonstrator off the showroom floor (plus ten others) for helper service on Cajon Pass and other steep grades in Southern California.

Other early buyers of the HD model included Pittsburgh & West Virginia (22) and the Akron Canton & Youngstown (6). But, as often happened, Pennsy's initial order for 12 with a repeat order of 26 represented the second largest delivery of any Fairbanks-Morse locomotive. In one small twist of irony, NYC subsidiary Indiana Harbor Belt, which served the EMD plant at McCook-LaGrange, bought 19 of the HD units that pulled many a new unit on its first trip into the outside world. The 96 orders for the H-20-44 over seven years was eclipsed by the nine-year production of the traditional road switcher models, H-15-44 (35 units) and its 1,600-hp successor, H-16-44, which sold 209 U.S., 58 Canadian, and 32 Mexican units. Like Alco and Baldwin, F-M offered a C-C model (H-16-66), but sold only 59 in three years. But the resilient technical staff at Fairbanks-Morse had one more surprise up its sleeves.

Figure 8.7. Ohio's 160-mile Akron Canton & Youngstown ran all of its mainline trains
behind Fairbanks-Morse power, as illustrated at top by an eastbounder getting orders as
it crosses Detroit Toledo & Ironton rails at Columbus Grove in June 1960. The lead unit
is brutish No. 500 (H-20-44) displaying the famous F-M "lifted weight" herald on its nose.
At bottom, a side view of this train leaving Delphos depicts the relative sizes of the two
F-M models. No. 500 was delivered in 1948, and the H16-44 was three years younger.
J. Parker Lamb.

Figure 8.8. The Train Master road switcher represented a major leap in locomotive power that Fairbanks-Morse achieved from the high power-density of opposed-piston engines. Two silver-trucked demo units, TM-1 and TM-2, clad in a gaudy scheme of orange, red, and yellow, clump over a crossing diamond at Decatur, Illinois, in 1953. The lower silhouette of an eastbound Wabash train, gliding past snow-covered Central Illinois farms in December 1958, illustrates the Train Master's impressive bulk when compared with the smaller FA and F3 units. Louis A. Marre Collection (top), J. Parker Lamb.

With their most powerful C-Liner (12-cylinder, 2,400-hp engine) essentially dead after the arrival of the hood-unit era, the company decided in 1951 that a potential market existed for a supersized C-C road switcher, larger than Alco's RSD models. The H-24-66 model was soon on the drawing board, and in 1953 four demonstrators rolled out of the Beloit doors and started touring the nation. In order to divert attention away from EMD and Alco, the company realized its new unit needed a slick pitch from Madison Avenue, and so it hired George S. Cohan to come up with a name and sales campaign. He settled on the name Train Master, and thus the demo units were numbered TM-1 through TM-4. Once again, with the AAR convention in Atlantic City as the venue for their formal debut, the bulky machines (TM-1 and -2), clad in a circus wagon paint scheme of red, yellow, and black, were a center of attention.

As with the earlier H-20-44, there was an immediate response to the impressive performance by the demonstrators, and orders began flowing in. Delaware, Lackawanna & Western was especially anxious to get a machine that could handle both freight and commuter trains in the same day. It ordered 10 units as well as 6 H-16-44 units (B-B road switchers). The Virginian Railway, about to dieselize its nonelectrified operations, saw the Train Master as its primary road power and ordered 19 units, along with 6 H-16-44s for local runs. Out west, after running TM-3 and TM-4 through their paces, Southern Pacific immediately bought them and ordered 14 others. Coal hauling Reading Railroad, like Virginian, picked 17 Train Masters, and Central of New Jersey wanted 13 units because of their dual service capabilities.

But with all of the Train Master's good features, it was fighting an uphill battle for sales in a nearly saturated market, and thus total deliveries were only 127 units over four years. As it turned out, the burly TM was the last major locomotive project for Fairbanks-Morse, which had become involved in the crippling environment of a hostile takeover process in 1957. Sadly, all domestic production ceased in 1958, although some orders for Mexico were not completed until 1963 after a total production of 1,460 units. Although there was never any doubt about the efficiency of the opposed piston design, its compactness was as much of a curse as a blessing. The general problem with Fairbanks-Morse units was the marked difference (from units of any other builder) in their required maintenance procedures. The upper crankshaft of the O-P engine mandated special handling when workers were changing bearings, connecting rods, or power packs. Much to the chagrin of F-M, some roads removed the O-Ps and replaced them with engines from either EMD or Alco.

Electro-Motive Catches Up

Record-breaking production of EMD's passenger and freight cab units between 1939 and 1957 (over 7,300 units) allowed LaGrange the luxury of letting other builders probe niche markets, such as that represented by hood units in the immediate postwar period. By 1947, however, it was becoming clear that these versatile locomotives were resonating with the nation's railroads, so EMD began by offering what it termed a branch line model, the BL1, constructed in February 1948. Mechanically it was a typical F3 unit but with a more utilitarian car body, having an F-unit cab (slightly longer nose) with visibility in both directions. But its overall appearance was unique in American diesel annals due to its longitudinal support structure, which consisted of a main frame augmented with small sidewall trusses (similar to those on E and F units). These sidewall trusses were covered with a sheet metal cowling that formed curved contours along each side of the unit.

After the BL1 demonstrator returned to LaGrange, a number of modifications were made, including a heavier underframe (to handle stresses of MU connections) as well as a new throttle control. The production version (BL2) was a modest success with 58 units sold to nine roads, three of which (B&M, C&O, Rock Island) assigned them to commuter train

Figure 8.9. Scrambling to join the road-switcher revolution, EMD introduced its uniquely styled branch-line model (BL2) in 1948. Only 58 of these units were produced, of which the last was Monon No. 38, at Lafayette, Indiana, in 1965. In the lower photo, a pair of Monon locals at Crawfordsville, Indiana, in 1959 produced a seldom seen view of the BL2 model. Alco RS-2 No. 21 was built in 1947, 16 months before the BL2. J. Parker Lamb Collection.

Figure 8.10. ElectroMotive's No. 300, last of the GP7 demo units, heads an Illinois Central train southward through Brookhaven, Mississippi, in 1950. C. W. Witbeck, Louis Saillard Collection.

service by ordering steam generators in the nose housing that exhausted through the center post of the windshield. Clearly this level of acceptance was not sufficient for EMD, as confirmed by a large pile of complaints from user lines as well as EMD's own service department. Max Ephraim, a young engineer at the time, recalled a three-hour meeting in 1948 with Cyrus Osborne, EMD general manager, and Gene Kettering, now chief engineer, in which the three men laboriously reviewed all major complaints. At the end of meeting, the BL2 project was officially declared dead, according to Ephraim.

In early 1949, Dilworth established an advance design team to work on a new limited potential, general-purpose locomotive. The colorful Dilworth recalled later his design philosophy this way: "In planning the GP locomotive I had two dreams. The first was to make a locomotive so ugly in appearance that no railroad would want it on the main line or anywhere near headquarters, but would want it as far as possible in the back country, where it could really do useful work. The second dream was to make it so simple in construction and so devoid of Christmas-tree ornaments and other whimsy that the price would be materially below our standard mainline freight locomotives." For example, he did not want an ammeter on the controls or the installation of automatic transition for the traction motors, and he certainly did not want dynamic brakes. In essence, Dilworth wanted to build a Model-T locomotive for America's railroads.

As the project moved along, Ephraim remembers that there was no styling as such. "We just covered the machinery with sheet metal." But there was one appearance feature about which Dilworth was insistent. He forbade the use of the then popular large number boards used on the streamlined noses of E and F units. Finally, he decided to incorporate angular end panels on the hoods and put the numbers at the top so they were clearly visible on an approaching locomotive. The model designation GP7 reflected simply the type of locomo-

Figure 8.11. In 1950
Nashville Chattanooga &
St. Louis received a half-
dozen lightweight GP7s
(with AAR Type A trucks)
that were intended for yard
and passenger service.
Delivered as No. 705 in
NC&StL's red-yellow color
scheme, this unit became
No. 1705 with the L&N
takeover in 1957. Its original
paint was still intact when
photographed at
McMinnville, Tennessee,
seven years after the
merger. J. Parker Lamb.

tive and the number series of cab units then in production. Incidentally, EMD vice president Nelson Dezendorf objected strenuously to the casual translation of the letters GP into Geep (pronounced jeep), even issuing a formal edict to company personnel prohibiting any such reference. Of course, such internal edicts had no effect on the legions of train watchers, commentators, and locomotive chroniclers.

Three GP7 demo units (Nos. 100, 200, and 300) were completed in late 1949 and were dispatched throughout the nation. Number 100 went north to C&NW, Milwaukee Road, and Soo Line, where crews suggested that the production models be winterized as had been done on Great Northern's F3s, which were modified with heated cowls around intake and exhaust vents. Demonstrator No. 200 worked westward from Salt Lake City to California, while No. 300, the first locomotive constructed at EMD's Cleveland plant, operated over lines south and east of St. Louis. This was the largest demonstration effort by any EMD unit since the famous 1939 tour by the FTs. The three demonstrators were sold to the Chicago & North Western in 1950. Following 20 years of good service, they were rebuilt in 1970 and operated for 10 more years. After retirement, the initial unit was restored to its original high-hood appearance and is now operated occasionally at the Illinois Railway Museum at Union, Illinois.

General response to the GP7 was the exact opposite of that to the BL model. It was an unusual case of complete resonance between builder and users. However, the Model-T simplicity that Dilworth dreamed about dissolved quickly, as virtually all users wanted improved transition controls, an ammeter on the control stand, improved cab heaters, larger fuel tanks, and even dynamic braking. The unique hood-top bulges for dynamic brake resistors are credited to engineer John Stanley, who finally grew tired of hearing his colleagues say, "There's no place to put it." EMD also found that, faced with a bidirectional mainline locomotive, railroads were divided about which end should be the front. Although the company had recommended that the short hood lead, many eastern roads opted for the improved crew safety afforded by a long end forward and were joined by the West's Great Northern. The height of the short hood became a subject of interest in the late 1950s when Phelps-Dodge mining company, which ran its units long-end first, ordered

Figure 8.12. Southern Pacific's lines east of El Paso, operated under the Texas & New Orleans banner, carried locomotive numbers less than 1000. EMD GP9 No. 246 (1954), shown at Lafayette, Louisiana, in 1956, is clad in the colorful "black widow" paint scheme of red, yellow, silver, and black. The large-diameter oscillating headlights used a rotating reflector and one large bulb. C. W. Witbeck, David Price Collection.

Figure 8.13. Eastern lines such as B&O, PRR, and NYC generally ran Geeps with the engine compartment shielding the crew. Brightly striped GP9 No. 3414 and mate lead the *Cincinnatian* northward from Dayton, Ohio, in October 1956. Enlarged fuel tanks pushed air reservoirs to the roof. J. Parker Lamb.

Figure 8.14. Electro-Motive's C-C trucked SD7 and SD9 models provided much greater tractive effort than Geeps with the same horsepower. A pair of 1955-vintage SD9s, augmented by an SW1200, head south from Peoria in June 1958 with a Chicago & Illinois Midland train. J. Parker Lamb.

Figure 8.15. Missouri Pacific found EMD's GP18 model to its liking and built up the nation's largest fleet (150 units), including 100 that used Type B trucks and GE traction motors from retired Alco FA/FB units. A late afternoon sun high-lights a pair of these solid blue units on a northbound train approaching Austin, Texas, in May 1964. J. Parker Lamb.

some low-hood units so the engineer could have better rearward visibility of his train on the walls of the deep open-pit mines. Later models would see increasing use of a forward-facing low nose for improved visibility at high speed

The evolution of the EMD's Geep models followed that of its F units into the -9 series at 1,750 hp (B-B) as the 567B engine gave way to the 567C (with 35 more rpm). With its powerful influence on the locomotive market, EMD's two late-starting general-purpose hood-type locomotives garnered some 6,216 sales in their first decade of production (including 170 cabless booster models). Of this total, 2,615 were GP7s built before 1954. Not surprisingly, the EMD colossus, though starting last, had sprinted to the head of the pack. Commentator David Morgan, who had characterized the *Zephyr*'s success with the comment "EMC had put on long pants," used a similar clothing metaphor for the Geep, suggesting that the diesel had donned dungarees. The hood unit was not a show pony like a streamlined cab model but a strong, everyday workhorse for all types of service. Three years after its B-B road switcher was introduced, EMD followed the lead of Alco's RSD models and introduced C-C versions of its own. Calling them special duty (SD) units, they were essentially elongated versions of the more popular Geep line. The SD7 and SD9 models sold 659 units between 1952 and 1959, slightly over 10 percent of the B-B versions.

Beginning in December 1957, EMD's famous V-16 567 engine reached its zenith in power output (without supercharging) when the 1,800-hp line of road switchers was introduced (GP18, SD18). They would be the standard power for four years, and they sold moderately well in a crowded market (350 GP18s and 54 SD18s), but in July 1958 the company jumped into the horsepower race that was developing with other builders (initiated by F-M's Train Master). It added a turbocharger to its 567D1 engine to produce a 2,400-hp package (567D3), transforming the SD18 into an SD24. Soon thereafter, it gave the Geep a 200-hp boost, producing the supercharged GP20 that used the same car body as the GP18. It was somewhat surprising that EMD, which had watched as other builders sweated through teething problems with these high-speed machines, was unable to avoid its own painful period of turbocharger breakdowns due to broken rotors and failed bearings.

The period between 1955 and 1960 saw the failure of two smaller builders and the emergence of EMD as a virtual monopoly, just as Baldwin Locomotive Works had been in the early days of the twentieth century. But the American system of capitalism will not tolerate such a competitive vacuum for long. In the steam era, a group of smaller builders combined in 1905 to create a larger competitor, American Locomotive Company, but the diesel era a half-century later would see a different process take place, although the results would be similar.

Figure 9.1. General Electric surprised the rail industry in 1959 when it began demonstrating a pair of heavy-duty road switchers, Nos. 751 and 752. Designated as Model XP 24-1, the pair and their attending dynamometer car glide into the Illinois Central yard in Champaign en route to Chicago in June 1960. J. Parker Lamb.

CHAPTER 9

A MONOPOLIZED MARKET

The post–World War II lineup of locomotive builders suffered its first casualty in 1950 when the two-year production run of Lima-Hamilton ended after it was absorbed into Baldwin-Westinghouse to form Baldwin-Lima-Hamilton. This was followed in 1954 when B-L-H, after severing its ties to Westinghouse, announced that a low level of demand would force it to cut back on locomotive production. Three years later, an internal upheaval regarding a change in management at Fairbanks-Morse caused it to cease production. Against this was the emergence of EMD as a dominant producer of road switcher designs. The giant builder had initially ignored this sector until 1947, and then its first attempt at introducing a new model produced a rare misstep. But with the 1949 creation of the GP7, it had roared away from the pack, leaving only Alco with a significant piece of the market.

Senate hearings in 1955 attempted to identify any antitrust activities associated with the continued dominance of the GM subsidiary in the wake of wartime allocations of materials and locomotive construction by the War Production Board. Among the tabular data presented in the Senate proceedings are those shown in chart 9.1, which presents the evolution of road switcher production over a seven-year span by the four builders mentioned above (exclusive of Lima). The distributions show Alco with a commanding lead until 1949 when Geep production began. Also clear is that Baldwin and F-M, taken together, never represented more than 17 percent of annual production after 1949. Further evidence of EMD dominance was the breakdown of diesels in service on January 1, 1954: EMD was at 61.7 percent, Alco at 23.9 percent, Baldwin at 10.5 percent, and Fairbanks-Morse at only 3.9 percent.

As noted, a federal antitrust lawsuit was filed by the Justice Department following the Senate hearings, but there were never any verdicts rendered against General Motors or even any formal proceedings. However, almost any unbiased observer will conclude that the needed wartime production allocations had the effect of creating a favorable competitive environment for one of the nation's largest manufacturing companies, which parlayed this prize into market dominance during the postwar decade. But, as also mentioned, this market situation was basically unstable in a free enterprise environment, and thus we find in the decade after 1955 a major reordering of the competitors for diesel locomotive production.

One of the first signs of instability came with the 1953 announcement of Westinghouse's intention to discontinue production of electrical traction equipment. While this adversely affected both Baldwin and Fairbanks-Morse, it was not a direct cause for either company's

Chart 9.1. Market share of
road-switcher production
between 1947 and 1954
shows the emerging
dominance of EMD.
*Proceedings of U.S. Senate
Subcommittee Hearings.*

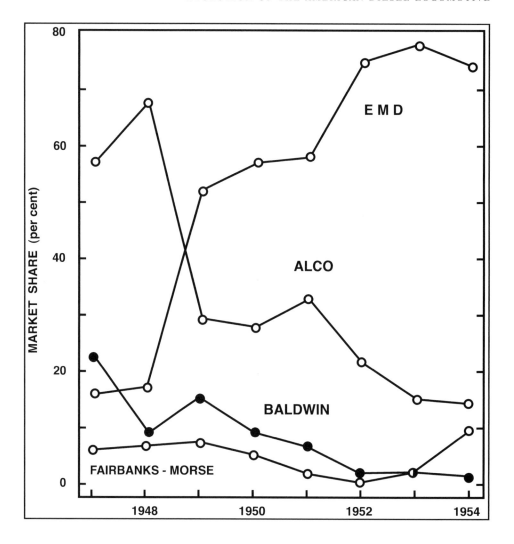

demise as a locomotive builder. Of more long-range effect was another announcement in
1953 that the joint-marketing agreement between Alco and General Electric would not be
renewed. Although from the outside this appeared to be an amicable partnership, there
had been many serious technical issues between the two engineering staffs. The electric
giant was clearly the instigator in the split, since it was planning some strategic decisions
and did not wish to be tied to Alco.

To Schenectady, however, the breakup was a crushing blow, since much of its locomotive technology consisted of GE proprietary designs. Worse yet, GE had gained enormous
engineering know-how during the life of the agreement. This background would provide a
significant advantage when the electric giant became an Alco competitor. In addition, the
cancellation meant that Alco could not hire another vendor to construct equivalent components. Although Alco shopped for other suppliers, including Westinghouse, it was finally
forced back to GE, knowing that it would pay more and yet not have access to GE's latest
developments.

Alco had a long history of diversification and, as a hedge on future growth in the U.S.
locomotive market, began in the 1950s to build more military equipment for the Korean
War as well as railroad rolling stock for the export sector. It also divested itself of financial
interests in related companies such as Montreal Locomotive Works (decreased to 17 percent) and General Steel Castings (liquidated all holdings). The character of its business
profile changed to include petroleum, nuclear, and marine applications, space technology, heavy machinery fabrication, and stationary power plants. Finally a new name was

tacked to the front door in April 1955 with the formation of Alco Products, Inc. Unfortunately for the new company, its U.S. locomotive sales, which had stood at $100 million in both 1952 and 1953, would never again approach that level.

A New but Old Player

Starting after World War II, the three major builders (Alco, Baldwin, and EMD), possessors of the world's largest and most advanced production facilities, began developing strong ties to the overseas locomotive market. During this period GE had also developed a line of export locomotives with lower heights and lighter weights than domestic units. These were marketed under the label Universal (meaning adaptable and versatile), with each model designated with U followed by the horsepower (in hundreds) and either B or C for its trucks. For example, a modified version of the well-known 70-ton industrial unit was offered as a U6B in overseas markets. In 1954, after fabricating an A-B-B-A demonstrator set, GE made clear its intention of moving away from its primary role as an export and industrial locomotive builder. Each B-B set was powered by a Cooper-Bessemer turbocharged engine, with one pair carrying 1,200-hp V-6 engines and the other 1,800-hp V-12s. After four years of road tests on the Erie Railroad, these units were fitted with 2,000-hp V-12s and designated UM20Bs. Sold to Union Pacific, they worked for four more years before being traded for new models. GE liked the solid performing Cooper-Bessemer engines and, on its next foray into mainline power, used the FDL-12 for ten U18B hood units sold to National Railways of Mexico in 1956.

What had transpired during this long incubation period was that GE had completely rewritten the locomotive design handbook. If it was going to compete in the class 1 locomotive market, it had to produce machinery that was as good as, or better than, what was currently available. Based on its direct contact with motive power managers on numerous roads, it had concluded that the most desirable characteristics at that time were simplicity and reliability and that the second usually followed the first. Design simplicity would reduce breakdowns and also cut maintenance costs, while thoroughly testing prototypes would increase reliability by completely debugging new equipment.

According to railroad shop people, one of the most desirable advances for any new locomotives was a cleaner air supply. Air is the lifeblood of a diesel locomotive, going into engine cylinders and circulating around traction motors, control cabinets, and other critical components. Dusty air filled with droplets of oil was a quick way to cause breakdowns. Some GE engineers put it this way: "We will have to design the air handling system and then build the locomotive around it." Thus the final design included a super efficient intake filter and a sealed supply system so that clean air areas were completely separated from contaminated portions of the unit. Another GE innovation was the self-draining radiator that could circulate the maximum amount of water during high-temperature operation and then automatically drain the coolant to a tank in subfreezing conditions. Potential wheel slip on the four-axle unit (with high-horsepower traction motors) was countered with a new detection and control system.

The heart of the locomotive is, of course, its engine, and many earlier locomotives were put into limited production with engines that were still in the testing stage. But that would not happen to the FDL-16, Cooper-Bessemer's four-cycle engine that produced 2,500 hp at 1,000 rpm. For ease of maintenance, the FDL power cylinders and their heads were an integral assembly, cutting the time for change-out by a factor of four. Jointly developed with GE engineers, the 45-degree Vee machine with its 9-inch bore and 10.5-inch stroke was as large as one could fit into a conventional car body. These specs showed a deep understanding of the need for growth potential in the railroad market, and they would prove to be a strong advantage for future GE units.

In mid-1959, three years after the U18B units crossed into Mexico, another pair of road

Figure 9.2. At top, the second set of General Electric U25B demo units, Nos. 2501–2504, rest between runs at Dearborn, Michigan, in June 1962, while below, Santa Fe's blue-and-yellow No. 1604, at Cleburne, Texas, in 1964, represents the early low-hood production configuration with a one-piece windshield. At bottom, the solid blue N&W No. 1900, the road's first U28B (July 1966), is in Chicago a month after delivery. Both Norfolk & Western and its future partner, Southern Railway, favored high-short hoods for improved crew protection. Louis A. Marre Collection (top), J. Parker Lamb Collection.

switchers rolled out of Erie's doors. They were designated as XP 24-1 (export demonstration, 2,400 hp) and completed 100,000 miles of demo runs. During this tour there was widespread speculation that these were not export models (too big and complex). The company confirmed this with its April 1960 announcement in that these were, in fact, its new U25B model. With this major step the industrial giant from Erie gave official notice that it intended to fill some of that production gap between GM and Alco. For the first time since its purchase by GM, the successor to Electro-Motive Corporation would face a competing company with financial resources and technology equal to its own. What GE still lacked was a reputation for durability of its locomotives. To accomplish this, it resorted to the same tactic as had EMC in 1939 when its FTs crisscrossed the United States on a successful demo tour. After GE's official announcement about its new locomotive, the original demonstrators (now numbered 751 and 752) went back on the road with a new purpose: to get attention and to generate sales. This time they ran 61,000 miles on a ten-month trip. But the business cycle was in a trough, making it difficult for the railroad industry to consider buying blocks of new locomotives. So the U25B units returned to the plant with nothing but stacks of impressive data from the dynamometer car, including a number of strong performances with heavy trains on grades, proving that GE's adhesion augmentation system (wheel slip control) was a winner.

The good impressions made on rail execs would persist, though, and after a second tour (74,000 miles on nine roads) was launched in early 1961 using the first four production units (Nos. 753–56), the results were much more favorable. By August, Union Pacific had ordered four units and Frisco had ordered eight (including the four demo models). In mid-1962, a third set of demonstrators (one with a low, short hood) was released and sales began to climb, with orders from ATSF, Erie, SP, and Wabash. By then the GE plant was producing 12 units per month, and GE's newcomer was proving to be a serious challenger for similar units from EMD and Alco. It was on the Rock Island in 1965 that the U25B supposedly received its nickname, U-Boat, and this time the source was operating crews rather than casual observers. By the time production ended in 1966, the largest U25B fleets were owned by NYC (70), SP (68), Pennsy (59), Rock Island, (39) and C&O (38). Indeed, when Penn-Central was formed in 1968 (including E-L and NH), its component lines owned a total of 178 U25Bs, almost 40 percent of the 476 produced.

Heavyweight Competition

By 1963 the American diesel market was completely reconstituted with GE as a far stronger competitor to EMD than Alco had ever been. For the next three years, there would be model-by-model competition from the three builders, with Alco finally fading away. The appearance of GE's U25B spurred EMD into production of a new B-B unit in July 1961. Its car body carried a new configuration, featuring a streamlined shroud partially covering the cab roof and extending backward over the hood to cover the fans and dynamic brake resistors. It was also the first model in which a low, short hood was standard. As the successor to the turbocharged GP20, its 2,250-hp 567D3 engine would have suggested the GP22 model designation. However, the advertising department took charge and decided on a sales campaign worthy of F-M's earlier Train Master. The ad theme became "A new locomotive with 30 improvements" over the GP20. Thus the designation was GP30, which began an unlikely numbers game between builders regarding locomotive designations. One would not normally associate such naming superficialities with the serious business of heavy machinery, but it continued for decades.

With thousands of its early units nearing the end of their operating careers, EMD began (with the GP20 and SD24) to emphasize a marketing strategy of unit reduction, which encompassed the trading of four or five old units for three new ones having the same capability. For example, four 1,500-hp units could be replaced by three GP20s, while three

Figure 9.3. Following the GP20's lack of success, EMD debuted a 2,250-hp B-B model in mid-1961. The GP30 featured an unusual sculpted cowling over the cab and engine hood. Nickel Plate No. 900, at Chicago in September 1963, displays the normal low-hood configuration (sans dynamic braking), while Southern No. 2537, with roof-mounted icicle-breaker (for tunnels) and a high-short hood, is ready for departure at Meridian, Mississippi, in December 1962. J. Parker Lamb Collection (both).

SD24s could replace five early units. In response, both Alco and GE were forced to adopt a similar process, but without as many older units available for trade, they often accepted almost anything on rails to make a sale. Many commentators viewed this period of replacement as the start of a second generation of dieselization, one characterized by widespread use of turbocharged engines for higher-powered locomotives. For chroniclers and observers, one interesting aspect of the trade-in craze was the use of trucks (and other components) from old units in the new ones. For example, GM&O bought its GP30s fitted with trucks from Alco FA/FB trade-ins.

General Electric's introduction of the 2,500-hp U-Boat gave it the temporary lead in the ongoing horsepower race that began as nation's railroads, now with better track and facing the need to move heavier trains at higher speeds, demanded a brawnier breed of diesel power. Alco and EMD had begun to respond with new models, but the U25B raised the bar to a new level. Thus in October 1963, almost 18 months after GE had ramped up its production to meet increasing orders and only two years after GP30 production began, EMD was forced to react quickly. It modified the 567D3, increasing rpm from 835 to 900, thereby producing 2,500 hp for its new GP35 model. This unit had no fancy styling as did the GP30; it had a spartan cab and nose with virtually no curved surfaces, only flat sheets

Figure 9.4. General Electric's introduction of the 2,500-hp U25B caused EMD to respond rapidly by beefing up its GP30 design to the same power level. A Memphis-bound Frisco train speeds through northern Alabama in June 1964 behind a newly delivered GP35 and an early high-hood U25B unit. The difference in heights of the two models is evident. In the lower photo, Western Pacific No. 3512, clad in a colorful orange and aluminum paint scheme, features a single-bulb, large-reflector headlight at Salt Lake City in June 1967. J. Parker Lamb Collection.

with angular connections. This would be the new EMD standard for decades. Production totals for the GP30/35 models were extremely impressive and indicated the readiness of American railroads to move rapidly into high-powered road locomotives. The GP30 sold 946 units in 28 months (including 40 cabless boosters for Union Pacific). During the following 27 months, the companion GP35 sold a whopping 1,250 units in the United States, 26 in Canada, and 57 in Mexico.

Both builders offered C-C versions of their 2,500-hp models in close succession, the U25C in September 1963 and the SD35 in June 1964. EMD soon realized it had squeezed all the power possible out of the venerable 567 configuration, and it began developing a new engine having a slightly larger bore (by 9/16 inch). The higher displacement gave this engine the designation as Model 645, as compared with GE's FDL power plant at 668 cubic inches. The next series of EMD locomotives (GP/SD40) would use the turbocharged 645E3 (900 rpm) producing 3,000 hp. The car body of the 40 series was similar to

Figure 9.5. In late 1963, Atlantic Coast Line began taking delivery of C-C locomotives from each of the major builders, so as to compare their performance in regular service between Waycross, Georgia, and Birmingham, Alabama. Shown here are examples of each unit during its first year of operation. At top is EMD's first production SD35, No. 1000, at the Thomas yard in Birmingham. Next, a pair of Alco C628 models, led by No. 2004, heads a train near Bessemer, Alabama, while at bottom, a General Electric U25C waits for an assignment at the Waycross shops. Note early wheel slip detectors attached to each axle. J. Parker Lamb, W. E. Mims (bottom).

Figure 9.6. Delaware & Hudson, long accustomed to pulling heavy trains across mountainous terrain, found the new C-C units to its liking. A longtime Alco customer, it bought an 18-unit fleet of C628s in 1964–65, including the No. 605 at Colonie shops, displaying its Model 251 engine and generator. However, with the rapid advancement of GE technology, the road elected to buy a 12-unit fleet of U30Cs in 1967 and an equal number of U33Cs in 1970. At the Wilkes-Barre, Pennsylvania, yard in October 1967, a trio of the new GEs prepares to couple to its train, while the Alco units will follow later with a coal drag. Jim Shaughnessy (top), J. Parker Lamb.

Figure 9.7. ElectroMotive's SD45 design of 1965 pushed unit power to 3,600 hp by using an elongated V-20 power plant. Production totals for this early design and its later variants were nearly 1900, although the extra long engine would later be the cause of many retirements. Three members of the SD45 family are shown here. From the top, the blue-and-white demonstrator leads a westbound Southern Pacific train into Hearne, Texas, and in November 1968. Next, a group of early Santa Fe units lug LA-bound tonnage across the Cajon summit. Electing to delete the dynamic brake housing, C&NW gave its SD45s a somewhat bare look, as typified in the lower photo by No. 907 heading an eastbound train near Ames, Iowa, in 1973. J. Parker Lamb (all).

Figure 9.8. Starting in the 1960s, General Electric began offering a wide range of power levels in its head-to-head competition with EMD. At Richmond, Virginia (top), is Seaboard Coast Line U33B No. 1742 riding on Blomberg trucks, while below is a diminutive Maine Central U18B, heading out of St. Johnsbury, Vermont, in June 1980. J. Parker Lamb Collection.

that of the 35s with the main difference in appearance being the number of large, roof-mounted exhaust fans at the rear of the hood. The 40s carried three fans, whereas the 35s had only two.

Soon GE displayed an agility in engineering and production that was revolutionary in the locomotive business. Only two months after the 3,000-hp EMD unit was unveiled, the Erie plant coaxed 300 more horsepower from its 25 series and offered the U28 series, which sold 148 B-B and 71 C-C units within a year (1966). What's more, early in 1967, the U30 series was unveiled, and GE was again even in the horsepower race. And this is how the rivalry would be played out until the 1980s, much like a race in which the frontrunner

Chart 9.2. Annual
locomotive production of
General Electric and
ElectroMotive between
1970 and 1990. Data from
Extra 2200 South, no. 125.

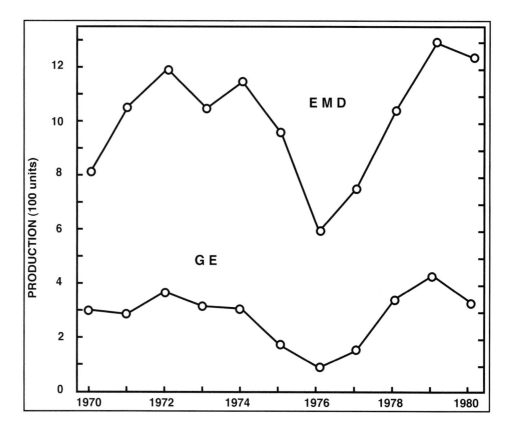

changes many times. One builder would initiate a new feature—usually more power—and then the other would catch up or even move ahead.

In particular, GE continued to show its ability to exploit the versatile FDL-16 engine design in the quest for more horses under the hood. It introduced the U33B/C in 1968 and the U36B/C in 1967/1971. The latter B-B designs pushed the four-motor configuration to its power limit for that period, but it was not as strong in the marketplace as EMD's much heavier SD45, which had been introduced in February 1965. Powered by the industry's first V-20 engine (645E3), this monster machine sold 1,312 copies within six and a half years, including 52 of a special passenger version with a steam generator. However, the engine's extremely long crankshaft proved to be an Achilles' heel, and many roads found the V-16 engine to be the most reliable configuration. Indeed, the V-20 design remains off-limits for locomotive use, even though tried again unsuccessfully in the 1990s.

The difficulties faced by GE as it fought to gain market share against EMD are illustrated in chart 9.2, which displays domestic production for both builders for the 1970s. The output from GE follows the same general distribution as that of EMD, with the proportion varying from 20 to 36 percent. Production for the two companies during this period was 12,891 (EMD) vs. 3,660 (GE). After a dismal year in 1976 when both companies hit low points, they rebounded in 1979 to hit peaks at 1,305 vs. 439. By 1981 both dropped precipitously due to an economic downturn. What happened after that was quite unexpected and will be reviewed in a following chapter that covers the second round of the great locomotive race between EMD and GE.

Alco's Final Push

During the period when its relationship with GE was unraveling, Alco was in the midst of a new engine development. This project had begun in 1949 when Alco president Percy Egbert assigned chief engineer Paul Vaughn to rebuild the basic 244 engine (9 × 10.5, 1,000

Figure 9.9. Alco's first Model 251-powered units were its RS series. A sample of these models includes (from top) a Southern Pacific consist near Austin, Texas, in 1967 led by an RS-11, assisted by an F3 and an RSD-5. In center, a high-nosed Northern Pacific RS-11, sporting both a vintage bell and modern radio antenna, was working ore docks at Superior, Wisconsin, in July 1966. At bottom is the behemoth of the line, the 2,400-hp RSD-15, represented by Cotton Belt No. 5158 at Pine Bluff, Arkansas, in 1965. The widespread visibility of Santa Fe's 50-unit fleet of these long-nosed units led to the nickname "alligators." J. Parker Lamb Collection.

Figure 9.10. Alco reached the apex of its diesel era with the Century Series units that were designed to achieve performance equality with its competitors at GE and EMD. From the top are a Seaboard Coast Line C420 leading an eastbound run-through train on the L&N near Biloxi, Mississippi, in 1968, followed by a New Haven C425 waiting in frigid conditions at Boston in February 1965. At the bottom is a newly delivered (October 1967) dark blue Chesapeake & Ohio C630 with high-adhesion trucks and the industry's first use of an alternator in lieu of a DC generator. These units had a short life on the EMD-friendly C&O. J. Parker Lamb Collection, Louis A. Marre Collection (center).

Figure 9.11. Washington State's Spokane Portland & Seattle, like Delaware & Hudson in the East, served as a haven for many generations of Alco units. Alas, creation of the Burlington Northern signaled the end of this relationship. Only a few months after BN operations began in mid-1970, an Oakland-bound manifest heads downgrade on Western Pacific rails toward Oroville, California, behind a CB&Q GP35, a Northern Pacific SD45, and a pair of SP&S Alcos (C636 and C424). To no one's surprise, these eastern-built orphans would not find a permanent home on the new BN. J. Parker Lamb.

rpm) to reflect the knowledge gained in the previous three years from service failures and intermediate fixes. Vaughn thought that a much stronger and more stable engine block was needed to combat crankshaft problems as well as to allow for future upgrades. The final design was approved in 1951, and thus it became the model 251. Its major improvement was the elimination of the 244's thermal regulation control system, which had produced hot spots that led to thermal expansion problems and ultimately to crankshaft fatigue failures. The latter problem was also attacked with a larger and stronger crankshaft. The application of an intercooler for the turbocharger outflow initiated the approach that was later applied to the last 244 models.

Learning from earlier mistakes, Alco did not rush the 251 into production as it did with the 244 model. Rather, it was put through an extensive laboratory and prototype testing program. A major part of the long test phase was a result of a serious rift between GE and Alco engineers over the design of a new supercharger manifold (delivery pipe) that would improve acceleration. The first 251 model installed in an operating locomotive was a small six-cylinder, inline, non-turbocharged version for the S-5 switcher, in order to gain operating experience before application to a more complex Vee-configuration.

In 1954 two 12-cylinder 251s were completed and installed in a pair of Lehigh Valley FA/FB units for a one-year test. Soon the final designs for the V-12 and V-16 versions were ready for production. In the meantime, car body designs for new mainline units were ready for an early 1956 announcement of Alco's 251-powered RS-line, which included six combinations of engine sizes and truck axles, all featuring angular front-cab panels. The new

Table 9.1.
Production of B-B Road Switchers

	Alco				GE	EMD		
	RS-11	RS-27	RS-32	RS-36	U25B	GP18	GP20	GP30
1960	25	1	—	—	—	134	101	—
1961	18	4	15	—	12	42	111	1
1962	—	21	20	25	69	119	42	459
1963	—	—	—	15	92	42	—	486
Totals	43	26	35	40	173	337	254	946
Builder totals			144		173		1537	

Source: Steinbrenner, *Alco: A Centennial Remembrance*, p. 404.

catalog encompassed two 1,800-hp B-B units (RS-11 and RS-36), two high-horsepower B-B units (2,000-hp RS-32 and 2,400-hp RS-27), and two C-C units (an 1,800-hp RSD-12 and a 2,400-hp RSD-15). This plethora of models represented an attempt to counter the offerings of the other two builders, but the results were not good news for Alco. Sales results for the B-B model competition for the four-year period (1960–63) are summarized in table 9.1.

Although the market shares for Alco and GE were essentially equal, the momentum for future development was clearly on the side of GE. Faced with these disappointing results, Alco management recognized in 1962 that it could either wind down domestic production and concentrate on further developing its thriving international business or make one more attempt to become a competitive force in the domestic market against two of the world's largest manufacturers. Not surprisingly, the determined company was unwilling to turn its back on Alco's heritage as a locomotive builder. Thus in 1962 it announced an improved line of 251-powered locomotives to be known as the Century series (C models), which would incorporate many of the design innovations introduced by the U25B, such as clean air in sealed cabinets, improved electronic controls, and better maintainability. The model numbering scheme consisted of three digits, the first being the number of traction motors (per truck) and the second referring to the horsepower (in hundreds).

The initial line contained three models: a 2,000-hp B-B, a 2,400-hp B-B, and a 2,400-hp C-C. However, the last model was never built, being soon replaced by a 2,750-hp C-C, which was a more direct competitor to other models. A summary of sales for the Century locomotives during Alco's final years is illustrated in table 9.2. As the data indicate, Alco gradually increased the power in its B-B units from 2,000 hp to 3,000 hp and its C-C units from 2,750 hp to 3,600 hp in order to stay competitive with other builders. The C630, for example, featured three major advances, the nation's first engine of this size, the first application of an alternator-rectifier in lieu of a DC generator, and the first application of a flexible, high-adhesion truck. But just six months after its introduction, EMD startled the industry with its 3,600-hp SD45 model, which pushed Alco to beef up the 251 engine further for its C636 model.

Having both the late-model 244 and the new 251 engine enabled Alco (like EMD) to offer a diversified rebuilding business in 1952. Its options included rework using a maximum number of useable parts or the replacement of an older engine with either of the later engines. While the bulk of a modest number of rebuildings were for first-generation units, the most interesting non-Alco rebuild was the 1964 repowering of Wabash's fleet of eight Train Masters with 251B engines rated at 2,450 hp, virtually the same as the original prime movers.

Considering its entire postwar production of road switcher models, those roads purchasing the largest fleets of Alco units were NYC (158), Southern Railway system (152), SAL (136), D&H (125), and Pennsy (121).

Although the Century units were arguably the company's finest machines, Alco sales slumped after 1966, reflecting widespread rumors that it would terminate domestic pro-

Table 9.2.
Production of Alco Century Series Locomotives

Model	1963	1964	1965	1966	1967	1968	Model Total
C420	6	45	36	24	12	8	131
C424	27	48	20	3	—	—	98
C425	—	41	38	12	—	—	91
C430	—	—	0	2	14	—	16
Total B-B	33	134	94	41	26	8	336
C628	4	41	72	30	13	26	186
C630	—	—	3	60	14	—	77
C636	—	—	—	—	3	31	34
Total C-C	4	41	75	90	30	57	297

Source: Steinbrenner, *Alco: A Centennial Remembrance,* p. 455.

duction. The company, which had suffered recurrent labor problems as well as multiple reorganizations and continuous financial pressures, was indeed moving inexorably toward an exit. It finally happened on the last day of 1969.

The path Alco followed during its final years was a familiar one in the business world. First, it had become the subsidiary of a larger company (Worthington Corporation) in December 1964. Three years later, the parent company merged to form Studebaker-Worthington, and the motive power business was divided into separate divisions, one of which was Locomotives and Engines. In 1968 a weakened S-W Corporation began to divest major assets. Fortunately, the Montreal Locomotive Works, a longtime affiliate and business ally, was able to purchase Alco's locomotive designs and worldwide licensing agreements. MLW continued production of selected models for a number of years, for use in Canada and overseas, until it was acquired by the Canadian firm J. Armand Bombardier Ltd. in 1975.

Figure 10.1. Two General Electric gas turbine–electric locomotives await their next runs at Cheyenne, Wyoming, in June 1968. All Union Pacific turbine units carried the same nose styling and similar car body designs, but they had differing wheel arrangements. Jim Hickey, J. Parker Lamb Collection.

CHAPTER 10

SPECIAL-PURPOSE DESIGNS

Although much more efficient than its steam-powered predecessor, the diesel-electric locomotive was not without its weaknesses. During the decade following World War II, two large western lines embarked on extended operating trials of several fundamentally different locomotive configurations. These derivative configurations were motivated by contemporary problems in areas such as electrical equipment failure in low-speed, upgrade service, overall fuel economy, and insufficient power in each diesel unit (leading to long strings of units). Not surprisingly, the two roads at the center of these experiments were Southern Pacific and Union Pacific, both of which have long stretches of challenging terrain. SP's constant struggle, from its inception, was to move tonnage over the Sierra Nevada (Roseville, California, to Sparks, Nevada) while UP's line between Omaha, Nebraska, and Ogden, Utah, traversed the Great Plains, the Continental Divide, and the Wasatch Range.

Gas Turbine Propulsion

The earliest effort involved replacement of a thermally efficient diesel engine with an equally efficient gas turbine that could operate with a much cheaper fuel than the traditional diesel. The gas turbine is a long, cylindrical machine whose major components are fastened to its main rotating shaft, which connects the compressor in the front with the power turbine at the rear. Between these two devices are the combustors, a group of smaller cylinders oriented longitudinally and arrayed on the periphery of the compressor. The incoming air, after being pressurized, passes through the combustors. It then flows through the turbine where its thermal and mechanical energy is transferred into rotational motion of the shaft, thus turning the compressor as well as a gear-connected electric generator.

Due to the extremely high combustion temperatures, gas turbines can easily operate on residual oil that requires preheating in order to flow through the fuel supply lines (much like an oil-burning steam locomotive). The choice of a fuel is much simpler because turbine combustion is continuous rather than periodic as in a diesel, since the power rotor requires a continuous supply of hot gas. The rotating elements are finely balanced so there is virtually no vibration, even though rpm can reach 10,000. These rotational engines can produce enormous levels of power within a relatively small space, but the large rpm values

Figure 10.2. Union Pacific's two-unit gas turbine locomotives, powered with a dozen traction motors, were originally rated at 8,500 hp. This was later increased to 10,000 hp, making them the most powerful propulsion machines ever to run on American rails. Seen at Cheyenne, Wyoming, in April 1967. J. Parker Lamb Collection.

limit severely the engine's ability to accelerate and decelerate quickly. Moreover, its fuel consumption is nearly constant, depending very little on how much work the locomotive is doing. The primary reason for this seemingly illogical condition is due to the fundamental configuration of a gas turbine wherein a large portion of the power generated by the turbine rotors goes to the compressor rotors. This is required to keep the engine running. Thus the shaft output going to the generator is only a small percentage of the total power produced. In some sense, it is merely a residual component.

In the aftermath of World War II, gas turbine technology was still wedded largely to the aviation field, and thus two of the three manufacturers of jet engines, GE and Westinghouse, were anxious to develop new markets for their machines. In 1948, GE constructed the first gas turbine–electric locomotive (GTEL) for mainline service. It featured a double-ended cab-type car body, enclosing a 4,500-hp turbine (3,900 at railhead) and riding on a B+B-B+B running gear. The solid black locomotive, numbered 101, was first demonstrated on the Pennsylvania and the Nickel Plate railroads. After a year and a half of mainline work in the West, it found a home on the Union Pacific, where it was given UP colors and the number 50. This unit was never sold to UP and eventually returned to the Erie plant, where it was repainted black and given back its original number.

With its strong performance on test runs, UP ordered ten more turbines almost immediately. When delivered in 1952, the new GTELs were virtually identical to the demonstrator except for the inclusion of a single cab. Attached to the rear of the units were small fuel tenders to help satisfy the turbine's insatiable thirst. To minimize use of the main turbine within the yard, these units were also fitted with small auxiliary diesel engines. The design was further improved in 1954 when 15 more units were delivered with a gallery walkway cut into the side of the cowl car body. These units also used larger fuel tenders (taken from retired 800-class 4-8-4s) and were usually MU-ed with an early B-B diesel unit as a reliability measure, although no GTELs were ever stranded without fuel.

The pinnacle of American GTEL technology was embodied in 30 GE units delivered to the UP in 1958–61. Housed in a pair of long cowl car bodies, the lead unit contained an 850-hp Cooper-Bessemer diesel for hostling and dynamic braking, while the 8,500-hp main turbine (7,000 hp on rail) rode within the second unit, with a fuel tender at rear. In 1964 power at rail was increased to 8,500 hp (10,000 hp gross) while some units were fitted with nose MU connections in order to operate as a 17,000-hp power block. Other units

Figure 10.3. Even though Westinghouse's twin-engine, 4,000-hp B-B-B-B gas turbine–electric demonstrator unit operated satisfactorily, its radical differences from diesel machinery were too much to overcome, and the *Blue Goose*, as it was nicknamed, never made a sale in its two-year existence. Seen in primer paint at the Baldwin plant in early 1950. David P. Morgan Library.

were fitted with traction motors beneath their fuel tenders. While the UP units, known as Big Blows for their jetlike noise, were specialized machines that excelled in high-speed mainline service, the gas turbine's inability to operate economically except at full load eventually doomed this diesel derivative. All were traded back to GE between 1963 and 1969 where trucks and other components were used on more advanced diesel models.

In parallel to the GE approach, Westinghouse-Baldwin gas turbine No. 4000 was also a test and demonstration vehicle for exploring performance in both freight and passenger service. It was about two years behind GE's No. 101, leaving East Pittsburgh in April 1950. However, it contrasted with GE's unit by using two relatively small turbine-generator sets (2,000 hp each) placed side by side within a 77-foot car body. The unusual nose was similar to the earlier Baldwin Sharknose units with a forward-slanting prow, and it rode on four B-trucks that, unlike other large units, did not have bolsters between pairs of trucks. Instead, the car body was allowed to slide transversely over the trucks, although constrained by a spring system. The result was an extremely smooth ride. Painted blue and gray with orange stripes on its pointed nose, it soon gained the nickname *Blue Goose*. Its demonstration tours included heavy freight service on lines around Pittsburgh (P&LE and Union railroads) and passenger runs on the Pennsy (on the mountainous lines in central Pennsylvania), on the MKT between Parsons, Kansas and Denison, Texas, and on C&NW's runs between Chicago and Duluth, Minnesota.

Its general performance was as expected, showing more power and speed than a diesel of equivalent horsepower. For example, the C&NW tests compared fuel usage with two EMD E7 units (four 1,000-hp diesels). On a round trip it consumed 3,600 gallons of Bunker C (residual oil) vs. the E7's 1,600 gallons of diesel. However, with the price differential, overall costs were nearly equal, $175 (turbine) vs. $176 (diesel), but lube oil costs were tilted toward the turbine, with the *Goose* needing only 180 gallons for its two power plants and the four diesels requiring 640 gallons. Although No. 4000 was demonstrated extensively for two years, there were no orders, and it was returned to Pittsburgh for eventual scrapping.

The American gas turbine experience shows once again how a successful engine in non-railroad applications may not be well suited to hauling trains. Basically, the gas turbine locomotive was something of a hybrid, needing on-board diesel assistance and presenting an entirely new technology for maintenance and repair forces. It reiterated the difficulty of utilizing two or more types of locomotives without incurring extra expense. Indeed, it was for that reason that some regional lines went so far as to use locomotives from a single builder.

Twin-Diesel Platforms

Despite the failure of Union Pacific GTELs to display an economically feasible advantage over conventional diesels, the road continued its quest for larger locomotive units that would allow it to increase average train speed across the vast reaches of the West. In the early 1960s UP's chief mechanical officer, D. S. Neuhart, became interested in the twin-engine concept, which had been dormant since the days of Baldwin's centipedes in 1945. He reasoned that the higher initial cost could be offset by lower maintenance expense over the life of the unit. Thus it was no surprise when UP approached the three diesel builders, commissioning them to build prototypes of twin-engine units that would generate at least 5,000 horsepower. All agreed to design such extra large machines.

EMD's 1963 entry was the DD35B, a giant booster unit consisting of two GP35 power systems mounted on a 70-foot platform supported by D-D running gear. EMD's advertisements suggested that, by attaching a GP35 unit to each end of the long unit, one could create reliable power blocks of 10,000 hp. After getting 30 of the boosters, UP ordered 15 more (with cabs) in 1965. These were designated as DD35As. Then, in 1969, the road ordered a second round of twin diesels, and the result was the 98-foot DD40AX, an enlarged version of the DD35A that incorporated two upgraded GP40 power systems producing 6,600 hp. The X designation for this unit indicates some experimental features of the design. In this case it was an advanced version of EMD's modular electric controls (see chapter 11). As the pinnacle of multi-engined diesel production, these 47 units stand as iconic counterparts of the 24 Big Boys (4-8-8-4s) of World War II. Union Pacific decided to call these behemoths *Centennials* in honor of the 100th anniversary of the Golden Spike ceremony at Promontory, Utah. Indeed, one of these rare units (No. 6936) has been kept in operating condition by UP and is used regularly on special excursion trains.

The initial design offered by GE (1963) was the 275-ton U50, similar in layout to the DD35A, with two U25B power assemblies on the B-B+B-B running gear salvaged from the GTELs. Like EMD, GE received a second order and responded with a more conventional 79-foot C-C unit (U50C), again using trucks from the last group of turbines. These designs had a similar appearance when viewed from the front, but their car bodies were quite different in layout. On the U50 machine, the radiators were at either end of the hood, while the U50C's cooling equipment was clustered in the center, as they were on the three EMD designs. UP bought 40 U50Cs starting in 1969. However, their short career of ten years was a result of a construction flaw, the use of aluminum wiring.

Alco's twin-engine design was the last to appear (June 1964) and the least successful. Using Century series design features, the 86-foot, 275-ton C855 also used the four -B running gear to carry two V-16 Model 251C engines, each rated at 2,750 hp. Its boxy car body was more utilitarian than any of its competitors, and despite its strong design specifications (137,000 pounds tractive effort), the A-B-A set was never duplicated, being scrapped after seven years of service. The only other railroad to indulge in this grand experiment was the Southern Pacific, which bought three each of the DD35B and U50 designs but never exhibited much enthusiasm for these complex units, although they operated satisfactorily for nearly ten years.

An overall assessment of the GM and GE designs suggests that, while they were operationally satisfactory, they could not pass the test of maintainability, despite the earlier conjectures of Neuhart. With two engines on-board, half of the locomotive sat idle while under repair. They were doomed more by their economics than by their technology. Indeed, the twin-engine designs were fundamentally flawed by one of the basic laws of

Figure 10.4. When Union Pacific invited designs for a 5,000-hp, twin-diesel locomotive in 1960, EMD responded with the mammoth DD35B booster unit that included two GP35 power assemblies. One of Southern Pacific's three acquisitions is seen here behind the road's latest B-B units (U25B and GP35) near San Antonio in 1965. Meanwhile, GE's competing model (U50) carried two U25B power assemblies on four B-trucks. At Council Bluffs, Iowa, in October 1964, this unit displays radiators at each end. J. Parker Lamb Collection.

Figure 10.5. The second round of UP twin-engine designs included an A-unit upgrade of the earlier EMD booster. At top, a westbound manifest leaves Cheyenne yard in June 1975 behind DD35A No. 80 and a group of cabless hood units. The pinnacle of two-diesel units was the 98-foot DD40X, which produced 6,600 hp from two GP40 power packages. Two of these Centennial units, led by No. 6909, maneuver a westbound train through the tight curves at Cajon Summit in November 1968. J. Parker Lamb.

Figure 10.6. GE switched to C-C trucks for its second twin-engine model. Although the U50C's car body was similar to that of its predecessor, a major difference was that the cooling modules were concentrated in the center. No. 5027 blasts through Columbus, Nebraska, with a westbound train in June 1973. J. Parker Lamb.

probability. Based on each line's operating statistics, any single-engine locomotive has a specific availability (probable percentage of time it is capable of accepting an assignment). However, by permanently mounting two such units together, the combination will have an availability equal to the product of the two individual availabilities. Thus two 90 percent units mounted together would have an overall availability of about 80 percent.

Diesel-Hydraulic Power

The third major attempt to improve the basic diesel-electric configuration was motivated by shortcomings in the performance of DC traction motors. When running for an extended time in an overload condition, these motors were likely to overheat and, if the insulation failed, to burn. The most drastic solution would be to discard this component, and thus attention was placed on nonelectrical drive systems.

With increasing use of heavy, rubber-tired trucks for industrial and mine service, there was widespread development of hydraulic couplings between engines and drive wheels,

Figure 10.7. The least successful of the two-engine models (and last to appear) was the 86-foot Alco C855, shown at Cheyenne in May 1965. The original A-B-A set, delivered in mid-1964, was never duplicated. H. N. Proctor, Louis A. Marre Collection

employing liquid turbines rather than solid shafts to transmit torque, hence the name *torque converter*. In operation, the power turbine would blast a swirling liquid onto the receiving turbine, causing the rpm of the receiver to be virtually the same as that of the input (thus a high efficiency). When applied to locomotives, such heavy-duty hydraulic transmissions were capable of delivering maximum engine power over a wide speed range, while ensuring that there would be no wheel slip or overheating in continuous low-speed service.

The original idea of hydraulic transmissions for diesel locomotives came from a 1925 theoretical study by the German engineer Alphonse Lepitz, who was later a member of Alco's technical staff. In the late 1920s, using Lepitz's concepts, the German Railway Office designed three sizes of standard drive trains (130, 360, and 550 hp) for application to self-propelled coaches and switchers. With widespread electrification of the German State Railways in the 1930s, diesel-hydraulic locomotives became standard in all nonelectrified territory. After the war, with its electrification destroyed, German rail systems increased the use of diesel-hydraulic units. Builders of the standard line of locomotives (switcher, freight, and passenger) were Mekydro and Voith for transmissions, while Maybach and Daimler-Benz furnished engines. Wheel arrangements included B-B, C, and D (the latter two being single-truck units).

The first American application of hydraulic drives for diesel-powered vehicles was the popular RDC (Rail Diesel Car) built by the Budd Company of Philadelphia, famous for the CB&Q *Zephyrs* of the 1930s. With the end of World War II, there was an explosive growth in auto and airline travel and a diminished need for railroad passenger service. In many cases, railroads were forced by regulatory bodies to continue operating commuter and branch line service, and they desperately needed a less expensive operating mode than the traditional train. Thus in 1948, the Budd Company, builder of a 1930s vintage rubber-tired railcar, sensed that there was a market for a modern version of the McKeen mechanical cars of 1904.

Figure 10.8. The 15-car RDC fleet of the British Columbia Railway (ex-Pacific Great Eastern) allowed it to offer regular service on the 465-mile stretch between Vancouver and Prince George. Here a southbound train (RDC-3 leading two RDC-1s) nears Squamish, British Columbia, in June 1978. J. Parker Lamb.

Led by Walter Dean, Budd's design team worked out details with the propulsion system manufacturers, and production began on two demonstrator cars that were completed in 1949. The RDC used an 85-foot stainless steel coach as a platform, with a small control panel in each vestibule that did not interfere with intercar movement. Beneath the floor were two 275-hp V-6s from Detroit Diesel, each driving a hydraulic torque converter modified from an Allison design used in the army's Patton battle tank of World War II. With all its axles powered, the RDC possessed an enormous 8.7 horsepower per ton, enough to allow it to accelerate uphill using only one engine.

Budd offered four interior configurations to cover a wide range of uses. The RDC-1 carried only passengers (capacity of 90), while the RDC-2 included a 17-foot baggage compartment plus seating for 70, and the RDC-3 added a 15-foot RPO section to the baggage area and thus reduced seating to 49. A no-rider version (RDC-4) carried only RPO and baggage and was 11 feet shorter than the other two due to its extra weight. The primary exterior feature of the RDC vehicles was a Vista-dome type of cowling that housed radiators and exhaust fans.

Eventually, 398 RDCs would be constructed (ending in 1962) for 23 American lines, 3 Canadian roads, and 6 foreign operators. The vast majority (60 percent) were RDC-1s, whereas only 3.5 percent were RDC-4s. Included in the total were 30 special "half RDCs" built for B&M's 108-car fleet, the nation's largest. Seating 94, the RDC-9 had only one power truck and no control space, thus serving as a trailer-booster behind a regular car. The longest RDC operation was the Western Pacific *Zephyrette*'s thrice-weekly, 927-mile run between Oakland and Salt Lake City, which required almost 21 hours to complete. For a few years Rock Island also operated an 875-mile *Choctaw Rocket* run between Memphis and Tucumcari, New Mexico. Later its western terminus was changed to Oklahoma City. Canada eventually accumulated the largest fleet (107) after VIA Rail became the national passenger train operator.

RDCs were also durable vehicles, with over half the original fleet still operating in 1987 (with ages ranging from 25 to 38). It is clear that, while this diesel derivative did not bring about an overall increase in U.S. rail passenger traffic, it went a long way toward keeping the rail sector a viable part of the American public transportation system for a longer time than many had speculated. It also confirmed the German successes in using diesel-hydraulic propulsion for certain applications.

Much less successful was the attempt to use diesel-hydraulic units for lightweight, low-slung passenger trains that were marketed by Pullman-Standard under the name *Train X*. Both NYC and New Haven operated these short-distance trains behind small power units (1,000 hp) constructed in 1956 by Baldwin-Lima-Hamilton in one of its last locomotive

projects. Three units (B-2 wheel arrangement) were built using a V-12 Maybach engine driving a Mekydro transmission, both of which were mounted atop the lead truck (15-foot wheelbase). The NYC's train was named *Xplorer*, and the New Haven used a unit at either end of its *Daniel Webster*. Not surprisingly, these units contained many flaws and were not reliable performers. Thus their future potential never materialized, and they were retired after four years.

Although both EMD's Canadian subsidiary and Baldwin had produced some low-powered diesel-hydraulic units for industrial switching in the early 1950s, most were never marketed. Thus the subject of diesel-hydraulics for heavy freight service was not actively pursued until D&RGW decided in mid-1960 to investigate the capabilities of the large units built by the Kraus-Maffei Company of Munich. Rio Grande was intrigued by the performance characteristics of diesel-hydraulics in low-speed running on steep grades where traction motors could burn out. After visits to Germany for discussions about the development of an American locomotive, it became clear to Rio Grande that, to meet the significant expense of designing and constructing a specialized unit for U.S. operations, the road needed a partner. So it turned to Southern Pacific, which also had plenty of mountainous trackage, to share the cost of six prototype units.

The final design included a cowl-type car body with turret cab and a roof tapered near the top (European clearances). Riding on C-C running gear designed by K-M, it was powered by a pair of Maybach V-16, 1,770-hp diesels with a separate turbocharger for each bank of cylinders. Voith transmissions drove each truck and also included a provision for hydrodynamic braking. Comparison with an EMD F7 shows that the K-Ms were 15 feet longer, but height and width were essentially the same. The engines were of typical German design, much smaller, lighter, and faster running than U.S. diesels. For example, the two MD870s had bore/stroke sizes of 7.3/7.9 inches compared with 9/10.5 for the Cooper-Bessemer FDL-16, while the respective rpm values were 1,585 vs. 1,000.

Initial trials in July 1961 on the 2.5 percent Semmering Incline in Austria involved tonnages up to 1,000 and plenty of curves and tunnels. The most severe test was a train restart on a 2.7 percent compensated grade with 955 tons trailing. A maximum of 80,000 pounds of drawbar force was recorded, and while the unit started slowly, it gradually regained normal speed. With these encouraging developments, the six units were shipped in late October 1961 to Houston to begin their shakedown runs on U.S. soil. The SP units began work immediately after being set up at the Hardy Street shops, while the Rio Grande K-Ms were moved in consist through Ft. Worth and Pueblo to the road's Denver shops for setup.

Early on, it became clear that the engines and transmissions were performing as advertised. After riding some K-M test trains in November 1961, *Trains* magazine editor David Morgan made these observations: "A pair of K-Ms, with a two-unit diesel helper at rear, tackled an 89-car, 3,967-ton train on the 2 to 2.4 percent grades west of Helper, Utah. The hydraulics ran satisfactorily at 10 mph for almost an hour, yet the transmission oil temperature rose only to 205 degrees Fahrenheit (10 degrees above normal but 43 degrees less than the limit). This performance was equal to that of four F-units with complete freedom from short-term traction motor limits." What D&RGW concluded was that one 4,000-hp hydraulic could replace two 1,500-hp diesels (maybe even two 1,750s) in either fast freight

(*Facing page*): Figure 10.9. Together the Denver & Rio Grande Western and Southern Pacific railroads purchased 21 diesel-hydraulic locomotives (4,000 hp, C-C) from Kraus-Maffei of Munich, West Germany, in 1961. In the top photo, one of the three Rio Grande units, later sold to SP, rests between runs at Denver in July 1963. Southern Pacific placed later orders to K-M for hydraulic-drive units that used a road-switcher car body. SP's No. 9014 at Oakland in November 1964, a few months after delivery, typifies the second design that encompassed 18 units. Louis A. Marre Collection (both), Alan Miller (bottom).

or drag service regardless of grade or altitude. The implication of unit reduction is that diesel-hydraulics decreased the deadweight of locomotives by as much as a million pounds per train.

But the German-designed machines were not without problems, most of which could be classified as a result of cultural differences in railroad practices between the two countries. The main difficulty was the European use of fully pneumatic engine controls rather than the electro-pneumatic system in North America wherein control information is transmitted electrically and activation is by means of air pressure. The all-pneumatic system is much slower to react to the engineer's commands, and breakdowns cannot be easily discovered with common electrical meters. The same types of control delays were experienced with hydrodynamic braking vs. traditional electrically based dynamic braking. There were also heating problems in trailing units grinding upgrade in tunnels. The K-M's air intakes were located too high, and the engine exhaust would include considerable unburned fuel that occasionally torched outside the locomotives. Most of these problems were resolved by retrofits, but after three years of constant shop tinkering, the Rio Grande realized that the K-M locomotives were too heavy on maintenance costs, and thus sold its units to the Southern Pacific, which was having a much better experience with integrating its German hydraulics.

With more extensive shop force than D&RGW, Southern Pacific could afford to establish a corps of diesel hydraulic specialists at Roseville to iron out problems. Moreover, SP saw enough positive performance that it began negotiations with Krauss-Maffei for the design and construction of a unit more compatible with U.S. equipment that retained all the inherent advantages. The result was two orders for eight 4,000-hp units that were delivered to Houston in 1961 and 1963. These used a modified Alco-type C-truck and a road switcher car body with a European-style tapered cab. By all accounts they performed effectively over Donner Pass, with three hydraulics replacing as many as eight F-units.

In an attempt to include even more American design practice in diesel hydraulics, SP also contracted with Alco to build three large units. This time the Voith transmissions were driven by two V-12 Model 251 engines that produced 4,000 hp at the rail. Designated DH-643 (C-trucks, 4,300 gross hp), the units resembled an enlarged and elongated C636 model. This Schenectady monster was almost 76 feet long and weighed 200 tons while the second-generation K-Ms were 10 feet shorter and 20 tons lighter. Performance of the Alcos between Roseville and Sparks was impressive with a tonnage rating of 3,850 between Sparks and Truckee, compared with 3,050 for the K-Ms, 2,850 for an EMD SD9, and 1,800 for an F7.

However, despite generally good performances from its hydraulics, when they reached the time that heavy repairs were needed, the road decided that any advantages they once possessed were no longer valid, as diesel electric technology had advanced rapidly during the interim. Thus all 24 units were retired between 1967 and 1973, ending the nation's experimentation with alternatives to traditional designs.

In an ironic twist of technology, the once dominant use of diesel-hydraulic power systems for propelling heavy machinery in quarries and earthmoving applications was replaced during the 1980s. Builders began switching over to the traditional railroad design, using a diesel-driven DC generator with traction motors on each wheel, giving these huge machines extreme flexibility and agility.

CHAPTER 11

NEW HEIGHTS FOR DIESEL POWER

Most observers peg the start of America's diesel development around 1920. Thus the year 1970 marked the approximate 50th anniversary of this transformation. By this time diesel locomotives were products of a mature technology but with even more growth potential in place. The two primary builders, having emerged from the economic shakeout of the 1960s, were ready to embark on a new phase of rapidly advancing technology. Earlier discussions mentioned the importance of enabling technologies that led to the development of the IC engines and DC propulsion machinery. However, during the 1960s and 1970s there were rapid advances in electronic technology that would permit another major leap in locomotive performance and reliability.

Most observers suggest that the beginning of widespread use of superchargers was the start of a second generation of power that replaced its predecessors on at least a 3 for 2 basis, while adding thousands of units for traffic growth. In contrast, the start of a third generation is not universally identified, but one can certainly make a case that these electronic advances represented the initial stages in the creation of a fundamentally new type of diesel-electric locomotive that finally emerged in the 1990s.

There are two major electrical functions in a modern locomotive. One is the power transmission process that transforms the mechanical output of the diesel into DC, which is passed to the traction motors that convert it back into the mechanical domain for propulsion. The other electrical component forms the control system for the transmission process. First perfected by GE's Hermann Lemp, these components modulate the connecting electrical circuitry between the generator and traction motors so as to optimize the mechanical output at the rail head for a given engine setting. Early control systems were generally composed of air-actuated (electro-pneumatic) relays and switches in cabinets that were ventilated but not sealed or environmentally controlled. Although continuous improvements were made, it was not until the solid-state revolution that microelectronic components (transistors and diodes) could be mounted on circuit boards. This enabled a complex web of wiring to be reduced to the size of a large envelope (circuit board) and placed within sealed modules that sat inside environmentally controlled cabinets, although there remained some mechanical elements. In addition to a significant increase in reliability, the use of modular controls meant that testing and repair were now virtually instantaneous. For example, after a malfunction, a technician could pull out an individual circuit board, test it with hand-carried instruments, and immediately replace it if necessary. Discarded boards could then be analyzed in a shop and either scrapped or repaired for reuse.

Figure 11.1. EMD's SD70MAC was the first widely produced AC-drive unit, due largely to the Burlington Northern's purchase of nearly 800 beginning in 1993. BN No. 9586, clad in the green-and-cream paint scheme of the road's Executive Train, leads a southbound coal train at Palmer Lake, Colorado, in 1997. J. Parker Lamb.

By the early 1960s it was clear to both builders that the DC generator would soon reach the limits of its capabilities. As horsepower increased, this component was taking up more and more valuable space inside the car body, and higher power levels were exacerbating the ever-present maintenance problems with brushes and commutators. Thus experiments began on the use of alternators to replace traditional generators, since they were smaller and less expensive. However, the added complexity was that AC technology had not yet progressed to the level that traction motors were feasible. However, the field of power electronics and solid-state devices was capable of producing reliable, large-capacity, silicon-diode rectifiers that could transfer the output of the alternator to the DC traction motors. In pursuit of a locomotive alternator between 1962 and 1965, EMD constructed a group of experimental units with AC–DC drive using B-B and C-C car bodies. This led to its November 1965 introduction of the GP40 model with alternators as standard equipment. During this period, GE had also come up with an alternator design, installing some on test models of its U28B in mid-1966. Soon alternators were standard on new models of both builders. But it would take over two decades for the remainder of the power transmission system to be converted to AC. As often happens with technological progress, the precursor capabilities are not ready when the needs are first identified.

Overlaying the change to alternators was EMD's announcement in February 1972 that its new locomotives would include a series of improvements known collectively as the "dash 2" package. New features included fully modularized controls on heavy-duty circuit boards and a sealed-pressurized electrical cabinet, but also improved trucks with greater flexibility for higher adhesion as well as a more sensitive wheel slip control system. Initially covered were its five primary models: GP/SD38 (2,000 hp), GP/SD40 (3,000 hp), and SD45 (3,600 hp). Each carried a new designation such as SD40-2. However, by 1977 GE was ready to introduce its own new line of power. Deciding to play on EMD's use of -2, GE changed all models to -7 (for 1977). Discarding the somewhat redundant U-notation, the U30B/C models became, for example, the B30-7 and C30-7. External appearances were unchanged except for minor details.

Another interesting development occurred in late 1976 regarding the tradeoff between total engine displacement and supercharging. During the difficult early years of turbocharger development, these high-speed machines were prone to breakdowns, and many roads preferred larger engines without augmented aspiration. However, as superchargers

Figure 11.2. At top, a pair of EMD's 2,000-hp GP38-2s lead an eastbound manifest along the SP's Sunset Route in West Texas in 1985. Below, a pair of Missouri Pacific GP40 units illustrate the evolution of this popular model. Lead unit, No. 904, a GP40-2 fitted with accessories such as an oscillating headlight, dynamic braking, and cab air conditioning, was formerly Western Pacific No. 349, whereas GP40 No. 631, carrying no such enhancements, was built in 1970 as Rock Island No. 4700 and came to the MP in 1984. Neither of these yellow units served for long with MP lettering. Shown at work in local service near Jewett, Texas, in 1986. J. Parker Lamb.

Figure 11.3. Over a period of 12 years and seven variants, EMD's SD40 family sold over 6,000 units in North America. Its ubiquitous presence is encapsulated by two Texas scenes. At top, Katy SD40-2 No. 635 races northward in morning light near Temple in 1988, displaying the long porches that resulted from its shared main frame with the SD45. Santa Fe, like many other roads, routinely operated mixed consists from both builders, as illustrated at bottom by a Houston-bound grain train near Rogers in 1985 behind an SD40-2 and a C30-7, both built in 1980. The "snoot" nose housing generally carried radio-control equipment for midtrain helpers. J. Parker Lamb.

became more reliable and ubiquitous, both builders found that numerous lines preferred a 12-cylinder turbocharged engine over an equally powerful, normally aspirated 16-cylinder machine. Their motivation was to avoid longer crankshafts and have fewer reciprocating parts to replace. As often happened, GE was the first to catch this wave with its U23B (2,250 hp) in mid-1968, while EMD answered with its GP39 (2,300 hp) a year later. Also during the early 1970s, EMD introduced its tunnel motor designs SD45T-2 (1972) and SD40T-2 (1974). Only two mountain roads, Southern Pacific and Rio Grande, opted for these units with lower air intakes that captured cooler air while the engine was running wide open in long tunnels. A total of 247 SD45T-2 and 310 SD40T-2 units were sold between 1972 and 1975.

The 1980s also saw a continuation of the pattern of locomotive improvements led by GE with responsive moves by EMD. Clearly the FDL engine design as well as GE's electrical/electronic technologies were tilting advancements more and more in Erie's favor. For example, GE's -7 line reached the pinnacle of its horsepower with the introduction of the C36-7 in 1978, followed two years later by the B36-7. Meanwhile, EMD had begun testing a GP40X model in December 1977. This unit was powered by an improved 3,500-hp 645F engine, and this led to the introduction of GP/SD50 models in May 1980 (GP) and May 1981. As noted earlier in chart 9.2, a general market slump at this time cut into production

Figure 11.4. General Electric was quick to take advantage of supercharger advancements to reduce engine displacement without decreasing power output. The resulting diesels were smaller and easier to maintain than their larger predecessors. Illustrating this evolution are two GE 12-cylinder units, U23B No. 557 (1976) and B30-7A No. 210 (1981), heading a UP local near Taylor, Texas, in 1988. Below, Southern Railway B23-7 No. 3982 (1985) illustrates the road's "crew protection" operation as it leads a southbound chemical train southward from Valdosta, Georgia, in 1980. J. Parker Lamb.

Figure 11.5. Special versions of the SD40-2 and SD45-2 models were the "tunnel motors" built for Southern Pacific and Rio Grande. In this design, the air intake grilles (at rear) were placed in a lower position in order to capture cooler air. The SP SD45T-2s are descending through Colfax, California, toward the Roseville yard in May 1972, while the D&RGW SD40T-2 is heading a westbound SP stack train near Seguin, Texas, after being displaced from Rocky Mountain service in the wake of the SP-Rio Grande merger in 1988. J. Parker Lamb.

figures for both builders, including for the 50 series. Thus only 278 (GP) and 361 (SD) units were sold, although the total was 50 percent larger than GE's numbers, 222 (B-36) and 214 (C36).

Most observers agreed that EMD had pushed the envelope too far with the 50 series, which generated numerous complaints about engine difficulties (connecting rod failures) and adhesion problems (new wheel slip controls). Later analyses suggested that the root cause was excessive vibrations (due to higher rpm), which shook the entire car body and all its contents, causing numerous components to fail and eventually to be redesigned.

This experience prompted GM to make major changes in its design. Primarily, it decided to create a 710 cubic inch engine by extending the stroke of the 645 model from 10 to 11 inches, creating an engine whose displacement was slightly larger than the FDL (668). Despite this, GE again jumped ahead with its new -8 models, led by the B36-8 (October 1982), the C36-8 (March 1983), and the C39-8 (May 1983). These units, initially tested in October 1982, included GE's first use of microprocessors to control all electrical systems, including a new wheel slip control system.

The first EMD models with the new 710 engine were the 60 series initiated by the 3,800-hp SD60 in May 1984 and the companion GP60 in October 1984; both models incorporated a trio of microprocessors for control functions. While the SD60 was virtually the same in appearance as its predecessor, SD50, the B-B model carried a slightly less spartan appearance than the GP50, with more rounded corners on the cab and nose, although the side cowlings for dynamic brake resistors were squared off. All roads except ATSF opted for the standard cab, but Santa Fe's second GP60 order in 1990 specified the wide-nose configuration known as the North American Cab. These units were also given the red-and-silver warbonnet paint scheme made famous during the early years of the *Super Chief*.

To upgrade its aging SD-40-2 fleet used in coal train service, Burlington Northern decided in 1986 to acquire a large group of SD60s. However, instead of purchasing 100 units, it developed a unique power by the hour concept. An EMD subsidiary, Oakway Leasing, was created to serve as the lessor of the units, and the railroad paid a flat rate for power (kW-hr) generated by the locomotives while in normal service. The Oakway units were clad in a derivative of EMD's blue-and-white scheme used first with the SD45 demonstrator in 1964. The long-term Oakway-BN agreement initiated EMD as a major player in the locomotive leasing business. Its fleet of available units would gradually expand to include most of its later models (including demo units), some of which were painted in the leasing road's colors. Incidentally, BN also engaged in a similar long-term agreement with General Electric that included 100 B39-7s lettered for the LMX leasing subsidiary.

By far the most shocking development in the locomotive world for decades was an announcement by General Motors in late 1985. The company said it had made plans to gradually shut down the LaGrange assembly line after 49 years and some 30,000 locomotives. Although EMD's executive offices and some component assembly would remain in Illinois, most major assembly would be transferred to its Diesel Division subsidiary in London, Ontario, founded in 1950 to service the Canadian market. It was obvious that GM's strategic planning group, after witnessing EMD's once commanding production lead vanish in 1983 and considering its future profitability, had decided to adjust GM's allocation of resources to the locomotive field. In simple terms, GM's rail-related activities would be downsized. Clearly, the era of producing 1,000 units annually would never return. The locomotive market had completed a 35-year growth period following World War II, but during the previous 5 years it had been shrinking, with locomotive rosters of class 1 lines decreasing by 10 percent. Thus there were fewer customers due to railroad mergers, and there was no dominant builder. In addition, the labor situation in Canada was more attractive than in the Chicago area. The clearest message coming from this move was that GM did not consider building locomotives to be one of its primary enterprises. The long-term consequences of this strategic decision were borne out by later annual

Figure 11.6. In the late 1980s, both Santa Fe and Southern Pacific decided on high-horsepower B-Bs for transcontinental intermodal runs. Both roads bought heavily into the GP60 and B-39/B-40 models. Whereas SP opted for the conventional cab design, Santa Fe decided to call these units its Super Fleet and gave them wide-nose cabs along

with the red-and-silver warbonnet paint scheme used originally on the famous *Chief* passenger trains. SP built a fleet of 94 GEs and 195 GP60s, and Santa Fe bought 60 units of each model with this configuration. J. Parker Lamb Collection (ATSF), J. Parker Lamb.

reports indicating that GE's rail-related business represented 1.74 percent of its 2003 sales, whereas the comparable number of GM was 0.64 percent and the dollar amounts were about 2 to 1 for GE.

After most of the EMD move was completed in 1988, the London plant began expanding to meet its broader responsibilities as part of the new General Motors Locomotive Group (GMLG). However, an increase in orders in 1994–95 caused GMLG to begin contracting the final assembly and painting to various shops, including Conrail in Altoona, Super Steel Schenectady in Glenville, New York, a Bombardier facility in Mexico, and two Canadian shops in Alberta and Quebec. Indeed, it is somewhat ironic that, during this phase, the latest GM locomotive company resembled an enlarged and modernized version of Harold Hamilton's EMC operations of the 1930s.

GM's 3,800-hp SD60 model sold 1,142 units, fitted with both traditional and North American cabs, whereas the companion B-B unit was far behind at 486. The GP60 demonstrator unit ran for three years before any orders were received. This outcome represented the new reality of American railroading, namely, that the smaller B-truck configuration had finally reached its market limit. It would no longer be the bread-and-butter model of a locomotive line. Instead, the C-C unit had again become dominant, as it was in the mid-1960s when development began on the Power River Basin coal reserves. The resulting coal train fleets fueled a ten-year production spike of 6,500 C-C units of which 5,408 were SD40s (and its derivatives) and another 1,091 were U30C, U33C, and U36C models. As for the GP60, it was the final member of the famous Geep family (15 models) that began with Dilworth's general-purpose locomotive, the GP7.

Even though B-B units were becoming history, the horsepower race for DC-powered C-C units still had one lap to go. In 1990, Erie pushed the FDL to 4,000 hp and labeled its newest models Dash 8-40B/C. These units were offered in both the North American cab (-40CW) and the full cowl Canadian configuration (-40CM). The Dash 8 line proved to be popular, selling 1,037 units between 1990 and 1994. GM responded to this challenge by upping the power of its 710 engine to 4,000 hp with the SD70 introduced in April 1993. This unit featured 42-inch wheels and GM's new EM2000 microprocessor system of control. Then, in another of the familiar move/countermove scenarios, GE increased output to 4,100 hp and then to 4,400 in 1994 with its Dash 9-44C. This forced EMD to offer the 4,300-hp SD75 a year later. While this unit sold a respectable 283 units in four years, the competitive Dash 9 series was running away with the market, selling over 3,000 units during the decade starting in 1993. It was after 1993 that production from the Erie plant began to far outstrip that from GMLG, with annual totals generally twice that of GM. Clearly, this represented one of the greatest turnarounds in the history of American industry.

One of the most significant advances coming from the last group of models from both builders, aside from power increases, was the introduction of radial or guided trucks that included more flexibility, both in their use of sliding pads to replace a center bolster and their movable axles. There were now three degrees of freedom for the axle on either end of a truck. While all three axles could move both vertically and laterally, the end axles could also rotate about a vertical axis near the centerline of the truck. This allowed each bearing

---

(*Facing page*): Figure 11.7. EMD's SD50 model incorporated numerous design improvements over the SD40-2. To house dynamic brake equipment, the hood was lengthened by 5 feet and the 645F engine produced 500 additional horsepower. CSXT's all-gray No. 8502, shown at Cincinnati in November 1990, displays the SD50's long and uncluttered look, while an Oakway SD60 at Glendive, Montana, was still toiling away in BNSF coal service in August 2004, nearly two decades after delivery. In the lower photo, the combination of a Katy SD40-2, leading a pair of UPs (SD60 and SD50) on a coal train near Temple, Texas, in 1988 encompasses a 20-year evolution of EMD C-C units. Louis A. Marre Collection (top), J. Parker Lamb.

Figure 11.8. General Electric's C36-7 was the last model to use the smooth car body contours of the early U series. With the (-8) and Dash 8 models, a much more utilitarian car body design was employed. At top, Union Pacific C36-7 No. 9013, part of the last 60 built, races eastward over former Texas & Pacific rails near Sierra Blanca, Texas, in 1986. Nine years later, Dash 8-40C's Nos. 9268 and 9323 (1988–89) lead a pair of SD70M demonstrators with northbound coal empties at Taylor, Texas. J. Parker Lamb.

housing on one side of the truck to move ahead of the corresponding one on the other side. The desired result was a greater ability to follow track curvature, ensuring full adhesion at all times. As anticipated, differences in adhesion between rigid and flexible trucks turned out to be truly significant, with the latter delivering over 40 percent, almost twice that for a rigid configuration (see chart 1-6). The common name for these new designs was *hi-ad* (high adhesion).

While the idea of unit reduction had been a longtime marketing strategy for more powerful locomotives by both builders, the action taken by Union Pacific in 1999 pushed this concept to a level never before seen. Recognizing that it had accumulated many locomotive models of various ages through its own purchases as well as through its railroad acquisitions (MP, WP, MKT, and SP), UP realized that its repair and maintenance operations had grown more complex and expensive. Consequently, it decided to shed many older models (especially its 1,700 SD40s) and standardize on fewer and newer models. It chose a contemporary DC machine as its mainstay because it was less expensive than an AC unit, whose primary advantages appear in the low speed range. The result from Omaha

Figure 11.9. These two locomotives represent the pinnacle of DC-drive units, and both were responsible for a major change in their builders' fortunes. The SD70M model (upper) was largely responsible for saving EMD from collapse due to Union Pacific's 1999 agreement to purchase 1,000 over five years. Flag-waving No. 5078, at Corsicana, Texas, in October 2004, represents the enhanced version of a 1993 design that now fills 1,500 roster spots on UP. But a much rosier scenario was created by General Electric's comparable models, beginning with the Dash 8-40C in 1987 and evolving into the Dash 9-44CW, exemplified by BNSF No. 4001 at Sealy, Texas, in August 2004. This basic design and its successors proved to be the industry's most popular contemporary units and vaulted GE to a commanding lead as the nation's premier builder. J. Parker Lamb.

was the announcement of the largest single locomotive acquisition in American railroad history: 1,000 SD70 units spread over five years.

Legally the agreement between the railroad and the builder was a letter of intent specifying firm orders for the immediate future, with acquisition plans for succeeding years. It also included a stipulation that GMLG continue to make refinements related to fuel efficiency and governmental regulations for emissions. There was an added benefit (to all railroads) for this massive order. It would give GMLG, now in second place as a builder, enough guaranteed work to maintain its viability as a competent manufacturer. Most observers believe that this bold move was extremely beneficial for all sides. By 2004, Union Pacific had exceeded its five-year plan, having received 1,200 of these DC machines, and was continuing to place more orders. Thus the SD70M had become UP's new universal workhorse, following the role played by its enormous fleet of SD40–2s some 40 years earlier.

With no further power increases by either builder since 1995, it appears that the pinnacle of power for DC locomotives is around 4,400 hp, over 3,000 hp more than produced by the FT model of 1939. Or expressed another way, one traction motor of a Dash 9-40B produces 1,000 hp, equal to three-fourths of an FT unit's total power.

AC Propulsion Is Realized

The asynchronous, or AC induction, motor has many characteristics that recommend it for locomotive use. For example, it will not overheat no matter how long it operates at a low (or even zero) speed because there are no brushes or commutators involved. Indeed, there are no moving parts other than the armature, making it smaller and much simpler to repair and replace than a comparable DC machine. Moreover, these AC machines presented few wheel slip problems. When teamed with new radial trucks, these motors proved to be ideal for unit trains of coal and grain in mountainous territory, although they could haul at high speed when necessary. However, the major reason for their delayed introduction, some 30 years after alternators, was due to the design of a reliable system of inverting (converting) DC from the alternator-rectifier into a shaped-wave AC that was needed by the asynchronous machine. Furthermore, control systems for AC motors were more complex than those for DC machines.

The complexity of the inversion process can be explained by first noting that the output voltage of an alternator can be represented by a common sinusoidal variation as depicted in chart 11.1a. Progress through the cycle is measured in angular degrees so that the AC frequency is of no concern. The variation begins with a zero value and rises to a maximum positive value at 90 degrees before decreasing to zero again at 180 degrees. It then reverses polarity and increases negatively to a minimum value at 270 degrees before rising again to a zero value at 360 degrees, completing the cycle. Thus the polarity changes twice during each cycle as it crosses from positive to negative at 180 degrees and from negative to positive at 360 degrees. Common household electricity flows at a frequency of 60 Hertz (cycles per second). Thus the change in polarity occurs 120 times per second.

The first wave conditioning process needed for diesel-electric locomotives involves the transformation of all regions of negative polarity into positive values (*rectification*). Thus the negative region between 180 and 360 degrees is switched to the positive domain, producing a two-hump positive DC (solid line in chart 11.1b). The wave can also be rectified with negative polarity (dashed line). The humps indicate that the DC signals are unsteady, but this can be virtually eliminated by the use of an RC buffer circuit (resistance and capacitance elements). Therefore, the final output of the alternator-rectifier in a contemporary AC locomotive is the uniform DC (+ and – polarity) shown in chart 11.1c.

Early AC-powered electric locomotives required a set of on-board AC motors to drive DC generators. In 1959, Pennsy began experimenting with rectifiers using large vacuum tubes (Ignitrons) filled with mercury vapor. Their successful performance led to an order

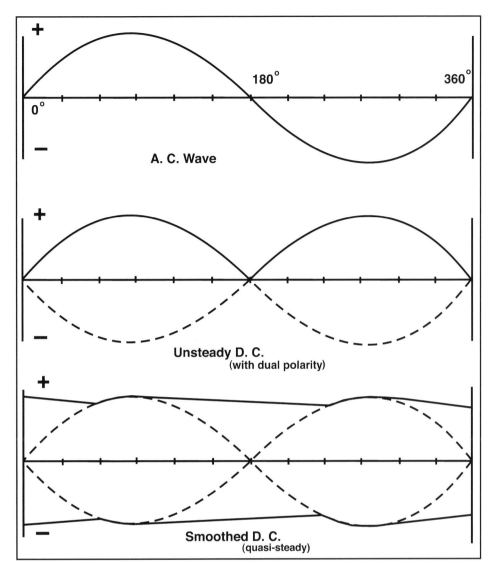

Chart 11.1. Schematic wave shapes illustrate rectification and smoothing of AC voltage signals between alternators and traction motors in modern DC-drive locomotives.

in 1963 for GE to construct 66 class E-44 locomotives (4,400 hp, C-C) that carried hood-type car bodies and running gear. By 1963 the use of power electronics for locomotives had emerged after the basic solid-state technology was developed. As noted in an earlier chapter, the demise of DC generators was made possible by the use of improved rectifiers using silicon diodes. Silicon is the key ingredient in many microelectronic devices because it can exhibit both electrical conductance and nonconductance, depending on the associated circuitry. Thus it is called a *semiconductor*. The last few E-44s were delivered with these smaller and simpler rectifiers that boosted their output to 5,000 hp.

Contemporary AC diesel locomotives required additional developments in power electronics. The process of converting steady DC to AC, known as *inversion*, requires another type of solid-state device similar to the diode. Known as a *thyrister*, it has the capability of being activated (fired) almost instantaneously to suppress a DC signal. In operation, one thyrister shuts off the positive DC signal near 180 degrees while, a few degrees later, another thyrister is fired almost simultaneously to pass the negative signal. Another switching process occurs near 360 degrees. This alternating gate effect, performed by a complex system of diodes, thyristers, and other electrical components, is known as wave-chopping and produces various forms of shaped-wave AC, two of which are shown in chart 11.2. Not only can these inverters produce many waveforms but they produce them at variable frequency, an important aspect in controlling an asynchronous motor.

Chart 11.2. Typical shaped-wave voltage signals used by AC induction motors of contemporary locomotives. These shapes are produced by three-phase AC from an alternator after passage through rectifier and inverter circuitry that uses solid-state components (diodes and thyristers). Personal communication, Prof. Mack Grady.

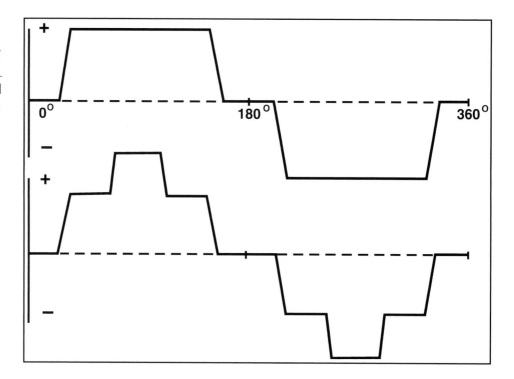

Chart 11.3. Schematic representation of three-phase AC, in which each wave (phase) is displaced from another by 120 degrees.

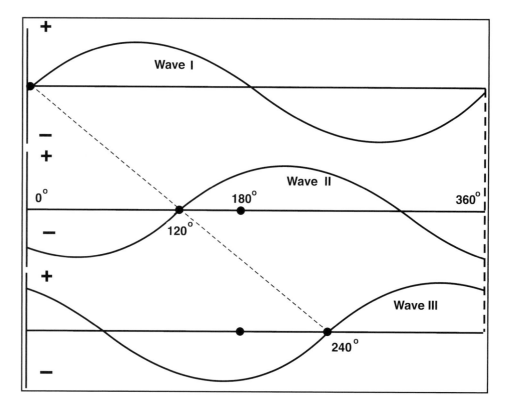

The foregoing discussion has considered only a single wire carrying electrical power to the motor. In practice, the oscillating character of AC allows it to be utilized in a three-wire (three-channel) network in which each AC wave in chart 11.1 is displaced from any other wave by one-third of a cycle (120 degrees). This is known as three-phase AC and gives induction machines a much greater power density than would otherwise be possible. Three-phase AC is often depicted on a single graph, but to simplify matters we show the

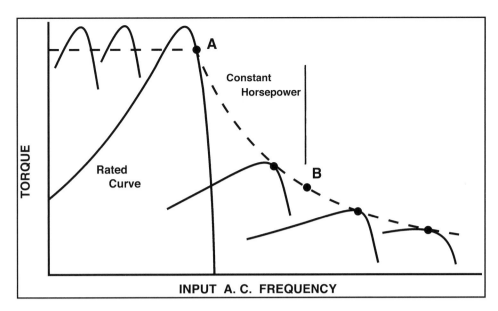

Chart 11.4. Performance characteristics of contemporary AC induction motors, illustrating the importance of input frequency. Primary features are a wide range of low-speed, high-torque output at low frequencies along with a constant-power region at higher frequencies. Internal curves depict constant voltage operation. Dashed line defines the envelope of motor operation. Bose, *Power Electronics and AC Drives.*

three signals (I, II, III) separately in chart 11.3. Each of these AC signals is rectified and inverted as discussed above.

While the details of AC traction motor performance are beyond the scope of the present narrative, it is nevertheless instructive to observe its different operating regimes, as presented in chart 11.4. This is a schematic depiction of the output torque of an induction motor. The dashed line is the overall operating envelope produced by many combinations of voltage and input frequency (examples are points A, B, C, E). At left is a low-frequency, high-torque region that is the AC traction motor's greatest attribute, allowing a locomotive to exert maximum drawbar pull over a wide range of low speeds.

Between points B and D is a region of constant horsepower best suited for high-speed running. The high-frequency region beyond D is of little use in locomotive operation, but it does show that the AC motor performance, at high frequencies, is equivalent to that of a DC-series motor (see chapter 1).

GM Locomotive Group

EMD began experimenting with an AC traction motor in the early 1980s, later fitting an ex-Amtrak SDP40F with a system using current source inverters from Siemens Transportation Group. This design did not perform satisfactorily, so in 1990 the company created two Amtrak F69PHACs with Siemens voltage source inverters and then ran extensive road tests. Strong performances in this testing led to the application of AC propulsion to a heavy freight locomotive. In 1991–92, four SD60MACs were built for testing on the Burlington Northern. With a successful performance tour by these units, it was only a matter of time before an SD70MAC was introduced. This occurred in December 1997 as the first of BN's 250-unit order rolled out and headed for the Power River Basin. This model proved to be a good seller for six years (1994–99), totaling 873 units, but then dropped to zero in 2001 and 2002, due in part to the outstanding performance of GE's competitive designs.

Unfortunately, some of GMLG's later efforts faltered as a result of engine availability. Thinking that bigger was better, the company made the first attempt to reach a 5,000-hp level when it unveiled the SD80MAC in July 1995. Like the SD45 some three decades earlier, it was powered by a V-20 diesel (710G3B) and featured a new 80-foot car body riding on 44-inch wheels with dynamic brake equipment located at the rear of the unit behind the massive angular radiators. Two demonstrators were constructed, followed by a 28-unit order for Conrail, while a 15-unit order for the Chicago & North Western was canceled

Figure 11.10. Using the same car body as the 5,000-hp SD80MAC, the SD90MAC was envisioned as a competitor for GE's AC6000CW, but was first marketed as an interim 4,300-hp unit that could be converted to 6,000 hp when a larger engine was available. In fact, it became an intermediate model between the original SD70MAC design and successive modifications that included new environmental controls. A downward view of No. 8213, denoted by UP as an SD9043MAC, leaves Topeka with westbound coal empties in October 2004, while a few months later, the modern SD70MAC model, represented by CSX No. 4718, is shown at Taylor, Texas, in UP coal service. J. Parker Lamb.

after the UP takeover. But even in the face of good performance, there was still a reluctance by most lines to embrace a V-20 configuration, especially when GE was offering nearly the same power with a smaller engine.

By the early 1990s, both builders were convinced that the next horsepower goal was 6,000, and they began scrambling to design a new AC unit, even as their sales staffs were discussing this significant advance as if it were imminent. In fact, neither company had even finished designing such a large power unit. Actual installation would stretch far into the future. For example, in late 1995, only three months after the SD80MAC appeared, GM announced the SD90MAC, which used the SD80MAC car body that enclosed a massive 1,010-cubic-inch V-16, four-cycle machine known as the 16V265H (or just 265H). This design, running at 1,000 rpm and using twin superchargers, broke EMD's tradition of using two-cycle diesels, going back to the Winton 201A. Unfortunately, when the 265H was tested during the preproduction period, its performance was grossly inadequate, forcing the company to install a substitute engine, the 4,300-hp 710G3 (same as SD75). Anticipating a positive outcome with the 265H, GMLG labeled the early models (SD9043MAC) as convertible or upgradeable, implying that the larger engine would be installed as soon as available, which turned out to be late 1996. In the meantime, GM sold 410 convertible units (309 to Union Pacific, 61 to Canadian Pacific, and 40 to a leasing company), but none has ever been converted.

The first of eight preproduction SD90MACs was unveiled at a railroad trade show in September 1996, while another was painted in special colors to commemorate EMD's 75th anniversary. These early units, designated SD90MAC-H, then went to the Department of Transportation Pueblo Test Center for extensive scrutiny, followed by shakedown runs on the UP. In the spring of 1998, production began on UP's 14-unit order. However, once on the road, the large units experienced constant engine problems and were often out of action for weeks at a time for modifications. In June 1998, EMD completed a pair of Phase II models that sported an even boxier cab and nose than the earlier models (for better visibility). Forty of these SD90MAC-H IIs were constructed for UP and Canadian Pacific (four units in 2000).

General Electric

The Erie plant stayed on the sidelines during the early days of EMD's experimentation with AC propulsion, in much the same way as EMD had let other builders explore the road switcher concept in the mid-1940s. However, once it became clear that AC propulsion was the wave of the future, the agile GE design group fielded a new model only two years after the SD60MAC test units were completed. Erie's first attempt (1992) was similar to that of GM, namely, to take its extremely successful Dash 9-44CW model and convert it to the AC4400CW prototype No. 2000.

Like the Geep of 1949, this new AC unit was exactly what American railroads wanted at exactly the right time. Sales skyrocketed, from a single prototype in 1993, to 68 deliveries in 1994, and then reaching an astounding 547 units in 1995 before settling in at about 300 units per year until 1999. The eye-popping total was 2,362 units through 2002. One feature of GE's system that attracted favorable attention was its use of one inverter for each traction motor (axle) as compared with the GM-Siemens design that serves each truck by one large inverter. Operationally, this allowed crews of GE units to shut down one motor when there was trouble en route, whereas the GM locomotives would lose the entire truck if any one of its motors was inoperative. However, the single-truck inverter produced much more uniform wheel wear. Another aspect was the cooling requirement. Since the large inverters of the GM-Siemens design generate considerable heat, they must be surrounded by a special electronic coolant, while smaller GE inverters for each traction motor can be air-cooled.

GE's quest for a 6,000-hp unit was more successful than that of GMLG. Based on a 957-cubic-inch V-16 engine (7FDL16A) developed jointly with Duetz MWM of Mannheim, Germany, GE's first offering in this power range was the AC6000CW of which 14 preproduction prototypes were constructed in 1995. Ten of these went to Union Pacific, and three were bought by CSX. These units were characterized by their extremely long fuel tanks (5,500 gallons), which occupied the entire underframe between the hi-ad trucks. Like GM, Erie constructed a number of convertible AC units to allow the production

Figure 11.11. To move coal trains on ex-D&RGW lines, Southern Pacific took delivery of 379 GE AC4400CW units between April and October 1995. Although most have long since been renumbered and repainted, SP No. 352 was still carrying both its original paint and number at Luling, Texas, in December 2005. GE's powerful AC6000CW model was three feet longer than its 4,400-hp brother, but it was still necessary to provide a much larger radiator housing with an unusual overhang at the rear of the unit, as demonstrated by UP No. 7559 at Taylor, Texas, in December 2005. J. Parker Lamb.

line for the 7FDL16A to ramp up. These used the same car body as the 6,000-hp unit but carried the FDL16 engine and were designated as AC6044CW models. Power-hungry UP purchased 106 convertibles along with 80 true 6,000-hp models, while CSX, with many mountainous routes for its coal traffic, quietly assembled the largest fleet (117) of AC6000CWs. The only other customer for this large locomotive has been an Australian iron-mining railroad.

It appears that, like earlier extra-large diesels, these powerful units may have a limited market. It is possible that both the cost and complexity of AC units have deterred many roads from building large fleets. With a price ratio of DC to AC units at two-thirds, the latter have found a secure niche only on road districts that require the largest levels of tractive effort, where the asynchronous traction motors can lug extremely high tonnage

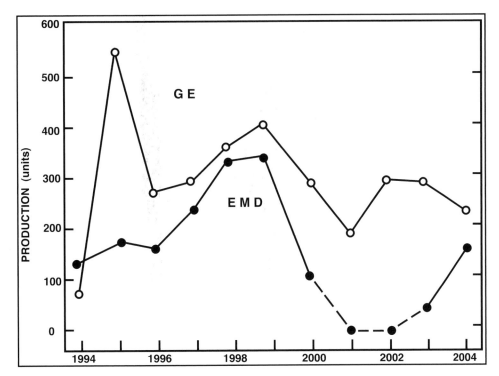

Chart 11.5. Production history of American AC-drive locomotives between 1994 and 2004. Data, *Extra 2200 South*, no. 123.

with ease. Three exceptions are Canadian Pacific, CSX, and Union Pacific, roads with the largest fleets of AC units. The early years of AC production are depicted in chart 11.5, illustrating GE's meteoric jump in 1995 followed by GM's pulling even around 1998–99. Both companies' production levels tapered off quickly in 2000 and 2001 as the shallow market for these high-tech machines became saturated. While GE rebounded in 2002, GM's output flatlined for 2001 and 2002, a bitter blow for the company that began the AC revolution. This disappointment merely indicated that major changes lay ahead.

Figure 12.1. Since introducing the Dash 8-40CW model in 1990, General Electric has kept its car body configuration virtually unchanged for both DC and AC models. Its latest unit, the low-emission ES44AC with the new V-12 GEVO engine, continues this practice. No. 5805 featured the revised BNSF logo as it worked in Powder River Basin coal service at Forsyth, Montana, in November 2005. Beth Krueger.

CHAPTER 12

RECENT DEVELOPMENTS

Environmental Challenges

During the half-century after World War II, the proliferation of diesel engines for American transportation (primarily highway trucks, construction vehicles, barge towboats, and locomotives) produced a significant public health concern regarding their emissions. Despite its many performance advantages, the diesel, like other combustion-based devices, releases both particulates (soot) and chemical toxins into the atmosphere. As federal government oversight has increased, more stringent regulatory limits have been introduced periodically. The U.S. Environmental Protection Agency began studying diesel locomotive emissions and their contribution to air pollution in 1977, although one of the first environmental aspects to be tackled by railroads was not any of those mentioned above. Rather, it was exhaust noise. An EPA regulation in the late 1970s required new locomotives to include stack mufflers after 1980.

Understanding the origin of diesel pollutants requires some awareness about the chemical aspects of both the fuel and the transformation processes in burning. Diesel fuel is a heavy product (low volatility) of the distillation of crude oil, a mixture of hydrocarbon liquids. It also contains trace amounts of sulfur as well as more carbon than gasoline (a lighter product). Furthermore, almost two-thirds of the air entering the combustion chamber is composed of nitrogen, a nonparticipant in the high temperature combustion process dominated by oxygen, which reacts with all components of the fuel and air to produce the flame and heat release needed to drive a diesel engine. Thus the products of combustion include carbon dioxide (CO_2), carbon monoxide (CO), sulfur dioxide (SO_2) from the fuel, and oxides of nitrogen (NO_x) from air. These were the principal pollutants identified in the 1977 government study.

Following a 1985 EPA report that recommended regulations for highway truck emissions be extended to locomotives, a 1990 amendment to the Clean Air Act set a five-year period for development of final regulations, a deadline later extended to mid-December 1997. As is usually necessary, the physical implementation of these new standards was a gradual process. For example, Tier Zero standards applied to all units built between 1973 and 2000 and were to be implemented at the time of the unit's next major overhaul. Tier One regulations were effective for new locomotives built through 2004, and they focused on reducing NO_x emissions from about 18.1 grams per kW-hour to about 8.7 grams, while

Tier Two levels covered new locomotives starting in 2005. These reduced NO_x to 7.4 grams per kW-hour and also lowered particulate levels.

One of the first approaches to pollutant reduction was to employ electronic fuel injection in new engines. This has been a common feature of automotive engines for many years, and both GE and GM began implementing this proven technology in the early 1990s. More recent technical approaches include a four-degree retardation of fuel injection that reduces NO_x by 28 to 35 percent with only a 1 or 2 percent increase in fuel consumption. Since the normal injection point is near top dead center (when the piston reaches its topmost position), this amount of retardation means that the crankshaft will rotate past TDC and will have begun its downward motion when injection begins. Another promising approach is to cool the exhaust gas manifold in order to reduce NO_x without affecting fuel economy. With a large amount of cooling capacity available in recent designs, this technology was relatively easy to implement in contemporary units.

To comply with these environmental limits, units of the same wheel arrangement and age are often grouped together for road duty, thus ensuring that controls and traction motors are compatible. Therefore, units with newer systems are not held back by coupling them with older models. Another practice is to slightly de-rate the highest horsepower diesels by 200 to 300 hp so as to decrease output of emissions. Units also can be shut down rather than being allowed to idle between runs. To do this, it is usually necessary to install an APU-driven heater for oil and coolants along with controls to operate it. The APU (40-hp diesel) emits only about 9 percent of the pollutants that come from an idling locomotive.

General Electric

As has often happened during GE's ascent to the leadership of locomotive development, its approach to emissions control has been to develop an entirely new engine design, with built-in growth potential, rather than to modify its veteran FDL16A power plant. The new GEVO model was a V-12 configuration that maintained a 4,400-hp output through additional supercharging, producing an immediate decrease in the volume of exhaust gas (and pollutants) by almost one-fourth. In tests, the new engine yielded a reduction of 40 percent in particulates and 3 percent in fuel consumption. The GEVO was installed in two new environmentally friendly models designated as ES44DC/AC (ES represents Evolution Series). Erie produced AC test units for Union Pacific and BNSF in 2003 and DC units for Norfolk Southern in 2004. These preproduction tests fulfilled GE's plan to have 50 units run for a year (i.e., 50 years of locomotive operating experience) before the start of production in 2005. Such a massive period of road testing is now deemed to be necessary in light of the high cost of these new locomotive models. Fortunately, the early performances of the ES designs were extremely strong, producing large orders from major railroads.

General Motors Locomotive Group

Electro-Motive met the environmental challenges by extensively upgrading its latest SD70M family into low-emissions models designated as SD70M-2 (DC) and SD70ACe, where the e indicates an enhanced capability. The car body for these units was essentially the same as that used on the SD90MAC-H II model, featuring a new cab with improved visibility and a vertical nose shape for improved collision protection. The new engine, designated 710G3C-T2, was reworked for emissions control while increasing power to the SD75 level (4,300 hp). Among the many improvements were a doubling of service

Figure 12.2. CSX received 20 EMD SD70ACe test units in May 2004, typified by No. 4848 waiting at Jacksonville between runs in August 2005. One of first production units, UP No. 8329, was already in coal service at Austin, Texas, in December 2005. The new unit's car body is essentially the same as that used with the SD90MAC-H II test units. Bill McCoy (top), J. Parker Lamb.

intervals to 184 days, segregation of piping and electrical layouts for ease of maintenance (piping on one side of the frame and electrical on the other), simplification of electrical circuitry (e.g., reducing 50 inverter-control circuit boards to only 2), and an on-board diagnostic computer program. The new engine reduced peak cylinder pressure by 15 percent so as to decrease engine wear and help meet Tier Two regulations. Finally, control cab ergonomics were improved by discarding desktop controls and returning to the traditional control stand, albeit with a worktable and computer monitor in front of the engineer.

By mid-2004, four demonstrator units were at work, including two in high-altitude testing on UP and others working KCS coal trains out of Kansas City over its mountainous lines in Kansas and Arkansas. Thirty additional units were constructed for a national demonstration program in 2005. Both performed well and led to early orders for over 100 units from a number of major lines (including UP, NS, and CSX) as well as a western regional, Montana RailLink.

In the meantime, there was speculation that its troubled parent had put EMD up for sale. By November 2004, GM confirmed that it was indeed in serious negotiations with a buyer, whose identity was revealed on January 12, 2005, when the sale of EMD to a pair of investment firms, Greenbriar Equity Group and Berkshire Partners, was formally announced. Greenbriar was a major investor in global transportation businesses, and Berkshire had previously held equity positions in midsize companies. The deal covered all aspects of EMD's railroad, marine, and stationary power plant businesses, from facilities in LaGrange and London, Ontario, to worldwide maintenance contracts.

Thus did EMD end its 74-year affiliation with General Motors. Its grand legacy was the fabrication of 58,000 units for 73 countries. The new owners announced immediately that the company's initials would remain EMD, although the third word would become *Diesel*.

New Technologies Emerge

With the switch to diesels after World War II, most roads made drastic cuts in their shop facilities and the considerable fixed costs they represented. Only a few railroads decided to invest in the conversion of their shops to diesel-oriented heavy rebuilding of their older units. Nine major lines engaged in modest levels of rebuilding, but as their rosters of older units decreased, these shops were gradually phased out. However, there were three large railroad-owned facilities that would continue with significant levels of activity until well into the 1980s. Illinois Central (Paducah, Kentucky), Santa Fe (Cleburne, Texas, and San Bernardino, California), and Southern Pacific (Sacramento) rebuilt large fleets of units, thereby extending their usefulness by a decade or more.

The eventual retirement of hundreds of older models spawned a national need for a ready supply of modern units for lease to class 1 lines during peak traffic periods. This situation led to the establishment of specialized locomotive leasing companies, which later expanded their fleets to include newly built units. Parallel to these developments was the spin-off of former railroad shops as independent companies serving the rebuilt and used locomotive markets. Some of them eventually began fabricating new locomotives using components from wrecked and partially scrapped units.

Among these independent fabrication facilities were the former Union Pacific shops at Boise, Idaho, which were acquired by construction giant Morrison-Knudsen. Precision National Corporation, another early rebuilder, acquired an old Ford Motor plant in Mt. Vernon, Illinois, for its construction activities. Morrison-Knudsen had an extensive background in building and operating rail lines for large projects, as well as a strong background in repairing diesel-powered equipment. It began locomotive rebuilding in the early 1970s, and during the ensuing three decades it engaged in fabrication of new units or performed major modifications to stock units. One of M-K's most ambitious projects was

Figure 12.3. In 1994, Morrison-Knudsen's Boise shops entered the market for small, urban-service units with the MK1500D. At top is an early morning view of Port Terminal Railroad No. 9608 at Houston's Pasadena yard in 1998. The lower view shows a 1997 descendant of the original M-K design from the successor company, Boise Locomotive Co. These units show the influence of the recent collaboration of EMD and Caterpillar. Pictured in November 2005 at Houston's Inglewood yard are CFEX No. 2040 and mate, both BLC GP20Ds. J. Parker Lamb (top), Ted Ferkenhoff.

the 1994 campaign to market a C-C demonstrator powered by a 5,000-hp Caterpillar diesel (Model 3612). Although this MK5000C model was upgraded in 1995 to include newer electronic technology, none of the six demonstrators was able to garner any sales to major systems. All were acquired by the coal-hauling Utah Railway in 2001.

A series of corporate restructurings changed the Boise facility to M-K Rail in 1994 and, three years later, to Boise Locomotive Company (BLC), a subsidiary of MotivePower International (MPI). In 1997 MPI was acquired by Wabtec Corporation, formerly the Westinghouse Air Brake Company. During this succession of ownership changes, BLC developed a line of low-powered, urban-service units with contemporary controls and emissions technology, thus capturing an important niche market that had been abandoned when GE and EMD discontinued production of such small models.

The use of Caterpillar engines by M-K Rail and MPI (BLC) was due to their lower emissions levels. Because they ran at much higher rpm levels than conventional locomotive engines, they produced significantly lower volumes of pollutants. M-K's first small units were 1,500-hp and 2,000-hp B-Bs riding on GP40-length main frames. Early buyers of the MK1500D were Houston's terminal roads (24 for Port Terminal Railroad and 8 for Houston Belt & Terminal) along with a pair of MK2000Ds for the New Orleans Public Belt.

After becoming independent, BLC improved its designs and entered a formal alliance with EMD and Caterpillar in support of production. Indeed, the current models are designated as EMD/BLC GP 15D and GP 20D. All BLC models share a high-visibility cab with short nose and low engine hood. Among the interesting features is the use of antifreeze protection for the engine, which can be shut down when not on duty in order to avoid the extensive pollution produced by idling for long periods. One of the major operators of these units is CIT Financial (CFEX), which owns 40 2,000-hp and 10 1,500-hp units.

During the 1990s, several small companies were engaged in studying further improvements in pollution levels that might be attained through the use of new technologies. From these efforts, three new technical approaches have emerged as important advances in industrial pollution control. The first concept was embodied in four switchers using liquefied natural gas (LNG), an extremely clean burning but expensive fuel. Built in 1994 by M-K at Boise with the same car body as its diesels, these units used a 2,000-hp Caterpillar diesel modified as a spark-ignited engine producing 1,200 hp. While tests on Union Pacific, Santa Fe, and Los Angeles switching roads were satisfactory, the lack of any governmental mandates at that time discouraged further development. In late 2005 the two ex-UP units were acquired by BNSF, which assigned all four to its subsidiary, Pacific Harbor Line in Los Angeles.

A more promising alternative was developed in 2001 by a Vancouver, British Columbia, builder. RailPower Technologies Corporation appropriated a recently introduced automotive propulsion system and produced a diesel-battery hybrid version of the traditional yard goat (switcher). Dubbed the Green Goat, it was built on a former GP9 main frame and used the Boise Locomotive car body (MK1500D). The original 2,000-hp (eq.) propulsion unit included a self-starting 130-hp diesel-driven alternator (90 kW), along with a bank of 320 rechargeable lead-acid batteries. In operation, all power is supplied by the batteries, which with the intermittent operation of switching service have a normal discharge time of four to five hours. When needed, they are recharged by the alternator-rectifier.

The prototype unit underwent testing in Canada and was then leased for a year to UP for tests in Roseville and Chicago. It later worked on the Pacific Harbor Line in Port of Los Angeles. In addition to yard duty, it performed successfully on a two-hour continuous run with a 1,500-ton train at speeds up to 62 mph. Later work on a full eight-hour shift of switching duty persuaded the company to increase the power of the recharge alternator to 200 kW. Comparing the Goat with conventional switchers showed fuel consumption was reduced by 50–80 percent and NO$_x$ by 90 percent (exceeding U.S. Tier Two require-

Figure 12.4. In 1994, both Santa Fe and Union Pacific leased the first demonstration units to operate on liquefied natural gas (LNG). Constructed by M-K in Boise using the same car body as its diesels, the four low-emissions units were aimed at usage in southern California. At top is ATSF No. 1200, at Los Angeles in 1994, while one of the UP units is below. BNSF began operating the entire group on the Los Angles Junction Railway late in 2005. J. Parker Lamb Collection.

Figure 12.5. Large, urban yards of Texas became homes to some of RailPower's first production Green Goat models. Examples include Railserve's yellow 2605 (top) and a bright orange BNSF unit that marries remote control to hybrid power. Switcher No. 1210 was being readied for service at Ft. Worth in February 2006. J. Parker Lamb Collection (top), Chris Palmieri.

ments). In parallel with the 2,000-hp hybrid, RailPower also produced a 1,000-hp unit built within a conventional switcher car body. It was dubbed the Green Kid. Both units showed in widespread demonstrations that they were ready for commercialization.

Parallel events proved that the introduction of hybrid drives was in complete synchronization with public policies on air pollution. Starting in 2001, a number of states and cities initiated formal programs for providing financial assistance for railroads and other diesel fleet operators who were capable of making significant reductions in emissions in the most severely affected areas. The Houston region, home of scores of chemical plants and oil refineries, was targeted by the Texas Emissions Reduction Program (TERP). By 2005 this program was up and running, with announcements of supporting grants to the three major roads in Houston (UP, BNSF, and KCS), all of which have ordered groups of improved Green Goats, officially known as RP20 models (or GG20). BNSF specified that some of its RP20s have no cabs, since they will be used only in a remote control mode.

In mid-2005 NS became the last of the major systems to test these hybrid power units, whose future seemed promising after RailPower's total orders had swelled to 175 units by October 2005. Recipients include big systems as well as industrial and contract switching companies such as Railserve, which operates 17 hybrids in Texas. Fabrication of Rail-Power's hybrid units occurs at the Alstom facility in Calgary, Alberta (ex-CNR backshop). Also in mid-2005 came promise of a more far-reaching use of the hybrid concept. GE's Transportation Group announced that it was developing a battery charging system for large road units that uses electric power from the dynamic braking system. The result would be an on-demand power boost from the battery bank for starting and for climbing grades.

Another new concept was announced in late 2005 when a U.S. builder unveiled its response to Union Pacific's 2004 request for a "gen-set" switcher, which also uses small, high-rpm diesels but is not a hybrid. National Railway Equipment, operator of the former Precision National plant at Mt. Vernon, Illinois, constructed a machine with two gen-set packages that use 700-hp Cummins diesels. The design allows independently controlled engines and traction motors to adjust to the immediate power demand by automatically shutting down and restarting. The stubby, center-cab prototype, built over the frame and trucks of an EMD MP15 switcher, performed brilliantly in its initial testing. Its overall capability was comparable to that of a 2,000-hp conventional unit, but with 80 percent less pollution along with a 40 percent fuel savings. Such promising results persuaded UP to order 60 such units from NRE in February 2006. These production machines include three gen-sets, providing a power range from 700 hp to 2,100 hp. The first 30 units were slated for yard service throughout California beginning in late 2006.

Green Goat originator RailPower was also included in the Union Pacific's rapid buildup of environmentally friendly power with an October 2005 order for 98 advanced hybrids, partially supported by a TERP grant of $81 million. This new fleet includes 18 models operating on a battery source (700 hp) plus two 590-hp gen-sets (RP20BH), while the 80 nonhybrid units (RP20BD) have two large gen-sets (1,880 hp total).

Figure 13.1. What better way to symbolize nearly two centuries of transportation progress since the Industrial Revolution than this juxtaposition of a white SD40 and a tethered white horse at DeQueen, Arkansas, in April 1989? Internal combustion power has permitted the pulling power of a single animal to be magnified many thousandfold. In this case the ratio is 3,000 to 1. J. Parker Lamb.

CHAPTER 13

THE DIESEL CENTURY
IN PERSPECTIVE

While the diesel-electric locomotive captured the nation's attention with the 1934 debut of Burlington's *Zephyr* and its daylong, flawless journey from Denver to Chicago, its development sprang from numerous collective efforts dating to the first decade of the twentieth century. However, by the year 2000 this remarkable machine had evolved into a source of power whose sophistication was unimaginable by the diesel pioneers and maybe even by such forward thinkers as Charles Kettering. Indeed, many will say that, even acknowledging the final heroics of American steam locomotives between 1940 and 1950, the twentieth century encompassed the period during which the diesel cycle was incubated and steadily developed into a reliable and widely used propulsion system for world railroads. Thus the term *Diesel Century* seems appropriate.

The horsepower race between builders was an expected part of the diesel locomotive's maturation. Clearly, Baldwin's 3,000-hp centipede of 1945 and Fairbanks-Morse's 2,400-hp Train Master of 1953 were far ahead of railroad needs of their periods. However, EMD's introduction of the 2,400-hp SD24 in 1959 was a more timely attempt to move locomotive power beyond the standard 1,500–1,800 hp level of the first generation of freight diesels. The immediate competition manifested by the sophisticated U25B of 1959 and the U25C of 1963 began a serious race toward 4,000 hp with GE leading the way, both in dates and unit power. EMD's SD45 (1965), with its supercharged V-20, set the bar at a new level, but GE's U36C and C36-7 soon showed that 20 cylinders weren't necessary. Indeed, a comparison of the SD45's 180 hp/cyl. with the latest GEVO 12-cylinder engine (366 hp/cyl.) illustrates the impressive results of three decades of technical progress.

The diesel-electric was responsible not only for a revolution in railroad operations but also for the demise of a large chunk of American heavy industry. Some of the leading machinery builders of the late nineteenth century, whose greatest achievements occurred during the frantic production demands of World War II, became victims of the changing paradigm. No longer could locomotive fabricators succeed by being experts in either mechanical or electrical propulsion, because the diesel-electric machine demanded a new breed of builders with expertise in both areas. Only two survived the challenges, and neither had been a participant in the steam era. As Churella's recent analyses have shown, the diesel locomotive demanded not the small-batch mode of production that was a steam era characteristic but the auto assembly-line model used by a company such as General Motors. It was a construction mode that demanded a high level of accuracy to ensure complete interchangeability of components, both small and large.

Figure 13.2. GM&O, a customer in the early days of Alco diesel production, completed its motive power fleet with Alco units after World War II because its delivery schedules were shorter than EMD's. However, by the time the FA/FB models were ready for replacement, the reputation of EMD far overshadowed that of a struggling Alco enterprise. Thus, like many other early dieselizers, the road chose the new GP30 model as its second-generation freight hauler. No. 501, only a few months old in late 1962, provides a stark contrast to a pair of road-weary Alcos in Meridian, Mississippi. J. Parker Lamb.

It is only natural for commentators and visionaries to attempt the risky task of constructing historical perspectives immediately following a series of events, and this has been true about the relative contributions of various diesel locomotive models. For example, the December 1970 issue of *Trains* included a listing of Ten Distinctive Diesels, compiled by longtime observer-analyst J. David Ingles. His criteria included solid performance, appearance, and popularity. Although not in numerical rank, the list included three models of the World War II period: EMD's FT and SW switcher line and Alco's RS-1. The second group, known for their appearance, featured the Alco PA and the Baldwin Sharknose cab units, while the third group represented hood units, both old and new. Units listed were Alco's C-420, EMD's GP18 and SD40, Fairbanks-Morse's best seller (B-B unit H-16-44), and GE's late entering blockbuster U25B.

The *Trains* issue of October 2002 included a new listing based on the broader view afforded by an additional three decades of development and production. For this compilation the selection keys included innovative design, a strong sales record, significant influence on railroad operations, and dominance against competitive designs. For 1939, the selection was EMD's early passenger units (starting with the E-3), and the following year it was the FT model. The wartime hero (1940) was Alco's pioneering S1 and S2 switchers, which, in the magazine's words, "chewed through the ranks of 0-6-0s and cleaned up city skies, too." By 1949 the appearance of the GP7 gave EMD a commanding lead in production, but by 1959 underdog GE had stormed the GM castle with its U25B model, which proved that the nation possessed a second competent diesel builder.

Responding to GE's aggressive growth, EMD pulled out all the stops with the introduc-

Figure 13.3. One of the iconic symbols of Electro-Motive's dominance in the diesel locomotive field for nearly seven decades was the unusual sideframe geometry of the Blomberg B-truck, which supported many thousands of units from LaGrange as they traveled throughout the world. Although the truck was well designed for its purpose, its artistic appearance has fascinated viewers since it emerged in 1939 as a component of the famous FT freight locomotive. J. Parker Lamb.

tion of its mammoth SD45 model in 1965, and by 1972 the smaller but more reliable SD40 family had become the most popular models in history. By 1987 GE's product quality and agile manufacturing had rocketed it into the production lead with its Dash 8-40C offering, while GM fought back in 1993 with the first production AC model, the SD70MAC, which proved the versatility of this new propulsion system. In the meantime, GE's more technically advanced Dash 8-44C and AC4400CW models became the market leaders and established the Erie-based company's dominance in a market that fortuitous wartime circumstances had dropped into the lap of industry giant General Motors. Thus the twentieth century ended with the two surviving builders having reversed their positions from the 1980 rankings.

An earlier section has included production comparisons (chart 9.2) during the first decade (1970–80) of head-to-head competition when GE struggled against the EMD juggernaut. The second part of that saga (1981–2002), illustrating the consequences of GM's downsizing and GE's technical and production superiority, is shown in chart 13.1. Although 1983 was the first year GE was able to best LaGrange's annual production (but just barely and only for a year), both companies slumped during the late 1980s. The year 1987 saw GE vault ahead again during a terrible two-year drought for GMLG with unbelievably low outputs of 93 and 161 units. But with 413 units in 1989, GM made its final appearance as the production leader. Largely on the basis of the phenomenal sales of its AC models in 1995, GE then pulled far ahead and cemented its leading position for the near future, although the end of the AC boom brought the two builders back together in 2001 and 2002. Another characterization of these two periods of locomotive production comes from average annual production. For the earlier period (1968–80), average annual output was 992 units for EMD and 281 for GE, whereas during the latter window (1981–2002) the comparable numbers were 338 (EMD) and 622 (GE). The downsizing of EMD

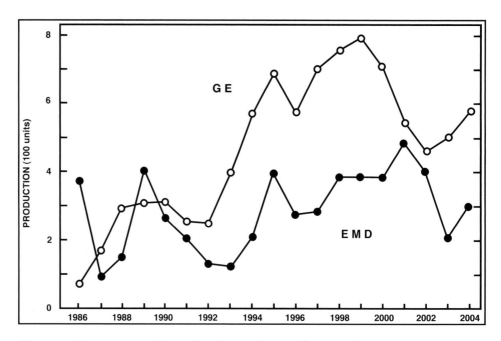

Chart 13.1. Continuation of annual production history shown in chart 9.2, showing gradual dominance of General Electric after 1992. Data, *Extra 2200 South*, nos. 123 and 126.

in 1985 indicated that the parent company was rethinking its corporate structure. The expected falloff in orders, especially for AC units, during the following years plunged the longtime industry leader to new depths that ended with its sale in 2005.

As the twenty-first century began, the ubiquitous presence of diesel power in the industrialized world had produced new challenges for mitigating environmental degradation. But the most serious problems for the first half of a new century are expected to involve the volatility of worldwide economics and its effect on the stability of fuel supplies. Of most serious concern is that these influences are controllable not by American technology or regulations but by the unpredictable nature of international political and ideological developments.

Earlier comments have mentioned one of the more amusing aspects of the Diesel Century, produced by generations of railroad commentators and photographers who had never seen a Pacific, a Mikado, or a Big Boy in their prime. They had matured during the last four decades of the twentieth century when the words *locomotive* and *diesel* were virtually synonymous. Thus their conversations were sprinkled with such phrases as "That southbound coal train has two Grensteins and a ghost." Translation: There were two BN SD70MACs (with the dark green and white colors ordered by BN chairman Gerald Grenstein) plus a ghostly white KCS EMD C-C unit. But such nicknames were not just for the later diesels. Even the early motorized coaches were often given nicknames by local observers. Common names included Doodlebug, Jitney, Scooter, and Puddle Jumper.

This volume closes by recalling the romantic prose of one of the most imaginative railroad writers of the twentieth century, David P. Morgan of *Trains* magazine. His picturesque phrases could transform trains, and especially locomotives, into something akin to living beings. In the original presentation, he refers to an accompanying photograph of a Western Pacific F-unit taken in the late 1960s. The author has placed Morgan's text beside another scene, recorded some ten years later, that confirms what his readers always knew: his verbal landscapes are virtually timeless.

"The Diesel Spirit," by David P. Morgan

Does a piece of machinery possess a soul, a lifelike spirit all its own? Of course not, some would say. Yet, as far as trainwatchers are concerned, that question was answered for all times years ago by that magnificent wonder of the mechanical age, the fire-eating, smoke-spouting steam locomotive—a machine that literally pulsed with dynamic life and, yes, human characteristics. Our question thus becomes more involved. Does a diesel locomotive possess a personality, a soul, a spirit? Back when shiny new diesels were replacing steam locomotives, the prose of the day was replete with references to "soulless diesels," "lifeless diesels," and "impersonal diesels." Glamorous and exciting—yes; lifelike—no.

Consider this scene on the Rio Grande Railroad at Minturn, Colorado, in June 1975 as an eastbound train begins the long climb through Tennessee Pass. Miles of tunnel running have dulled the diesels' once shiny coats with numerous layers of exhaust soot while clouds of sand have coated their trucks and running boards. Yet the classic contours of the EMD hoods are still etched in relief by a setting sun. Those V-16s in a dozen GM car bodies are chanting tales of endless canyon miles subdued, countless desert battles won. The confetti of youth has settled, time has brought esthetic appreciation, and the spirit of the diesel has emerged to reveal an endearing, no-nonsense workhorse—a steed that truly was alive.

Adapted from *Our GM Scrapbook*

Figure 13.4. The signal was red at the south end of Missouri Pacific's yard in Taylor, Texas, when this scene was recorded near sunset on a summer day in 1981. But as soon as the high green blinked on, the lead GE U23B came to life and guided the two EMD SD40s and the trailing MP manifest toward San Antonio. Clearly, it was impossible then to envision any scenario that, in less than 15 years, would propel General Electric into the position as America's premier builder of American diesel locomotives. J. Parker Lamb.

REFERENCES

Aldag, Robert A., Jr. "How Fairbanks-Morse Got into the Railroad Business." *Trains*, March 1987, 24–36; April 1987, 48–57.

——. "Culture Clash: Diesel vs. Tradition." In *Railroad History*, special issue, *The Diesel Revolution* (2000): 89–99.

Bose, B. K. *Power Electronics and AC Drives*. New York: Prentice Hall, 1986.

Brill, Debra. *History of the J. G. Brill Company*. Bloomington: Indiana University Press, 2001.

Brown, Charles A. "From Issl's Effort to Seaboard's Solution." *Trains*, May 1982, 38–45.

Burrows, Roger. "Here Come the Hybrids." *CTC Board, Railroads Illustrated*, July 2004, 18–23.

Churella, Albert J. "Business Strategies and Diesel Development." In *Railroad History*, special issue, *The Diesel Revolution* (2000): 23–37.

Crouse, Chuck. *Budd Co., The RDC Story*. Mineola, N.Y.: Weekend Chief, 1990.

Diesel Era. *The Revolutionary Diesel: EMC's FT*. Halifax, Penn.: Withers,1994.

Dolzall, Gary W., and Stephen F. Dolzall. *Diesels from Eddystone: The Story of Baldwin Diesel Locomotives*. Waukesha, Wisc.: Kalmbach, 1984.

Doughty, Geoffrey. *New York Central and the Train of the Future*. Lynchburg, Va.: TLC, 1997.

Ephraim, Max, Jr. "Martin Blomberg, Designer Extraordinaire." *Trains*, October 1994, 46–49.

——. "The Beloved Geep." *Trains*, June 1995, 44–53.

Extra 2200 South, nos. 123 (2002), 125 (2003), and 126 (2004). Blaine, Wash.: Iron Horse.

Graham-White, Sean. "GE AC4400CW (Data Sheet)." *Trains*, March 2004, 46–47.

Hamley, David H. "Those People Made Good Locomotives." *Trains*, December 1969, 28–43.

——. "Ingersoll-Rand: Catalyst of Dieselization." *Trains*, December 1970, 25–45.

Hamley, David H., and Raymond T. Corley. "About the Railcars Which (Unintentionally) Forecast Dieselization." *Trains*, November 1973, 36–49; December 1973, 40–48; January 1974, 26–28.

Hemphill, Mark W. "Locomotive of the Future: EMD's SD70ACe." *Trains*, September 2003, 38–43.

Hirsimaki, Eric F. *Lima, the History*. Edmonds, Wash.: Hundman, 1986.

Howell, John R., and Richard O. Buckius. *Engineering Thermodynamics*. 2nd ed. New York: McGraw-Hill, 1992.

Iczkowski, Mike. "Daybreak at La Grange." *Trains*, August 1976, 29–31.

Ingles, J. David. "A Little Change from LaGrange." *Trains*, April 1972, 16–17.

——. "Corner Grocery Store in the Diesel Market." *Trains*, June 1972, 26–29.

——. "Ugly Ducklings Disperse." *Trains*, November 1987, 45–49.

Kratville, William W. "Knife Noses and Portholes." *Trains*, July 1960, 30–39.

Lamb, J. Parker. "Supernovas of Steam." *Steam Glory, Classic Trains Special Volume* No. 2, November 2003, 104–10.

Lipetz, Alphonse I. "Transmission of Power on Oil-Electric Locomotives." *Mechanical Engineering*, August 1926, 797–897.

Lustig, David. "Doyle's House of Toys." *Trains*, February 2004, 29–31.

Marre, Louis A. *Diesel Locomotives: The First 50 Years.* Waukesha, Wisc.: Kalmbach, 1995.

Marre, Louis A., and Jerry A. Pinkepank. *The Contemporary Diesel Spotter's Guide.* Waukesha, Wisc.: Kalmbach, 1989.

Marre, Louis A., and Paul K. Withers. *The Contemporary Diesel Spotter's Guide.* Halifax, Penn.: Withers, 2000.

McDonnell, Greg. "The Other Diesel That Did It." *Trains,* August 1982, 42–51.

——. "GM's Comeback Kid: The SD70MAC." *Trains,* May 1994, 34–39.

——. *U Boats.* Erin, Ont.: Boston Mills, 1994.

——. "EMD's Big Boy." *Trains,* September 1997, 38–45.

——. *Field Guide to Modern Diesel Locomotives.* Waukesha, Wisc.: Kalmbach, 2002.

McGonigal, Robert S. "Max Ephraim Remembers." *Trains,* September 1997, 48–51.

Middleton, William D. *When the Steam Roads Electrified.* Waukesha, Wisc.: Kalmbach, 1974.

Morgan, David P. "Thrifty Glutton." *Trains & Travel,* January 1953, 25–28.

——. "The Semmering Story." *Trains,* October 1961, 43–49.

——. "A Weekend in the Rockies with the KM's." *Trains,* February 1962, 34–37.

——. "A Locomotive Is Born." *Trains,* September 1962, 18–24.

——. *Our GM Scrapbook.* Waukesha, Wisc.: Kalmbach, 1971.

Pinkepank, Jerry A. "Born at Beloit." *Trains,* November 1964, 36–49.

——. "On Behalf of Baldwin." *Trains,* December 1967, 22–30.

——. "Everyman's Diesel Primer." *Trains,* December 1970, 18–23.

Pinkepank, Jerry A., and George J. Sennhauser. "Those Red Diamond Diesels." *Trains,* November 1963, 26–37.

Rashid, Muhammad H. *Power Electronics: Circuits, Devices, and Applications.* Englewood Cliffs, N.J.: Prentice Hall, 1988.

Reutter, Mark. "The Great (Motive) Power Struggle: The Pennsylvania Railroad v. General Motors, 1935–49." *Railroad History* 170 (Spring 1994): 15–33.

Scharchburg, Richard P. *Under No Man's Shadow: The Life of Eugene W. Kettering.* Flint, Mich.: General Motors Institute Press, 1987.

Smith, Vernon L. "The Diesel—from D to L." *Trains,* April 1979, 22–29; May 1979, 44–51; June 1979, 44–49.

Steinbrenner, Richard T. *Alco: A Centennial Remembrance.* Warren, N.J.: On Track, 2003.

"Ten Diesel Locomotives That Most Changed Railroading." *Trains,* October 2002, 36–45.

U.S. Congress. Senate. Subcommittee on Antitrust and Monopoly. Hearing regarding Senate Resolution 61. 84th Cong., 1st sess., December 1955.

Van Hattem, Matt. "EMD SD70M (Data Sheet)." *Trains,* October 2003, 34–35.

INDEX

Note: Page numbers in italics indicate photographs and illustrations.

Abraham Lincoln, 44
Adams, E. E., 39
adhesion limits, *10*, 11, *111*, 138
aerodynamics, 40
aftercoolers, 93
air pollution, 30, 55, 159–60, 164–67
air supply, 109. *See also* aspiration; turbochargers
aircraft engines, 11
Akron Canton & Youngstown Railroad (ACY), *96*, 97
Alabama, Tennessee and Northern Railroad (ATN), 65
Alaska Railroad (ARR), 66
Alco: and the Black Marias, 84; and the Century series, *120*, *121*, 122–23, *123*; and competition, 112, *114*; and diesel-hydraulic propulsion, 136; and the DL series, 61–62, *62*; and early diesel development, 26, 30, 58–63; and export locomotives, 109; and Fairbanks-Morse units, 99; and freight trains, 52, 70–73; and GM&O, 170; and hydraulic transmissions, 132; and IC engines, 55; late diesel efforts, 118–21; and Lima-Hamilton, 88; and market share, 107–108, *108*; and partnerships, 108; and passenger units, 70–75; and production levels, 54, 69; and road switchers, 63–66, *65*, 91–92, 94, *100*, 108, 119, 121, 122, 170; and steam engines, 67–68; and switchers, 59, 60, *61*, 170; and twin-diesels, 128, 132
Alco Products, Inc., 109
"alligators," *119*
alternating current (AC) propulsion: adoption of, 150–53; described, 7–8, *151*, *152*, *153*; and EMD, 138; and GE, 155–57, *156*; and sales performance, 171–72
alternators, 122, 138
Alton & Southern Railroad (ALS), 91
Alton Railroad (CA), 15, 44
aluminum, 32
American Car and Foundry (ACF), 18
American Freedom Train, 74
American Heritage Foundation, 72
American Locomotive Company, 91–93, 105

American Society of Mechanical Engineers, 2, 22–23
Amtrak, 153
antitrust issues, 54
appearance features, 69–70, 91, 101–102, 143
armature, 6
Armco Steel, 33
aspiration: air injection techniques, 58; and Alco engines, 60; carburetors, 15–17; and fuel injection, 36, 160; and locomotive fundamentals, 9–11. *See also* superchargers; turbochargers
Association of American Railroads, 59
Atchison Topeka & Santa Fe (AT&SF): and Alco engines, 61–63, 72, 73, 75; "alligators," *119*; and Baldwin engines, 57, *57*, 58; and EMC-Pullman, 22; and Fairbanks-Morse engines, 85; and freight trains, 51–52; and FT models, 52, *53*, 78; and GP models, 143, *144*; and J. G. Brill, 20; and LNG units, 164, *165*; and McKeen units, 14; and passenger units, 44–46, *144–45*; and rebuilding business, 162; and the SD series, 116; and Super Fleet, *144*; and switchers, 47; and the Universal series, *110*, 111
Atlanta & St. Andrews Bay (A&SAB), 66
Atlanta Birmingham & Atlantic Railroad (AB&A), xii
Atlantic Coast Line Railroad (ACL), 114
Atlas Diesel Engine Company, 58
automatic control systems, 18
auxiliary power unit (APU), 160
aviation, 11, 40, 126
axle configurations, 11

babyface cab units, 81, 84
Baldwin, Matthias, 6, 107
Baldwin Locomotive Works: and advances in diesels, 169, 170; and Alco engines, 75; and "centipedes," 81; and DC propulsion, 8; and dieselization, 68; and early diesels, 55–58; and export locomotives, 109; and freight trains, 52; and gas turbine-electrics, 127; and the *Gulf Coast Rebel*, 82; and market share, 107, *108*; and NYC, 82; and production levels, 69, 77, 80–84;

Baldwin Locomotive Works (*continued*)
  and road switchers, 93–95, 96, 108; and sharknose units, 83;
    and VO models, 57–58; and wartime production, 54, 67; and
    Westinghouse engines, 32–33
Baldwin-Lima-Hamilton, 89, 95, 107, 133–35
Baldwin-Westinghouse, 107
ball bearings, 16, 60
Baltimore & Ohio Railroad (B&O): and Baldwin engines, 58, 84;
    and Brill cars, 19; and EMC engines, 50; and Fairbanks-Morse
    engines, 86; and FT models, 54; and Ingersoll-Rand units, 30;
    and J. G. Brill, 20; and passenger units, 45, 46; and road
    switchers, 103
Bangor & Aroostook Railroad (BAR), 69
Barriger, John, 85, 88
Batchelder, A. F., 16, 28, 29
batteries, 5, 32, 164–67, 166
"battleship" (M-2s), 15
Beardmore engines, 33
Beloit Wagon Works, 84
Belt Railway (Chicago), 60, 91
Benz, Carl, 1
Berkshire Partners, 162
Berlin Exhibition (1879), 6
Bethlehem Steel, 33
Big Blows, 127
Big Boys, 128
bi-power units, 30, 31
"Black Marias," 69, 84
Blomberg, Martin, 51
Blomberg trucks, 117, 171
Blue Bird, 25
Blue Goose, 127, 127
Blunt, James G., 59
Blunt-trucks, 59
Boise Locomotive Company (BLC), 164
Bombardier, 147
booster units, 44, 51, 70
Boston & Maine (B&M), 41, 42, 60, 91–92, 99, 133
bottom dead center (BDC), 2
Brake Mean Effective Pressure (BMEP), 10
branch lines, 13, 99–101, 100
Brill, Johann Georg, 18–20
Brill Company. *See* J. G. Brill Company
British Columbia Railway (BCOL), 133
Brown-Boveri Ltd., 60
Buchi, Alfred J., 60
Budd, Edward G., 13, 40, 40–41
Budd, Ralph, 39, 40
Budd Company, 132
Burlington Northern and Santa Fe (BNSF): and the ES series,
    160; and hybrids, 166, 167; and LNG units, 164, 165; and low-
    emission engines, 158; and the SD series, 146
Burlington Northern Railroad (BN), 121, 138, 143, 169
Burlington Railroad, 46, 68
Burlington-Rock Island Railroad (BRI), 50
Busch, Adolphus, 2
Busch-Sulzer Brothers Diesel Engine Company, 2

Cadillac automobiles, 35
Canada, 92, 96, 133

Canadian Locomotive Company (CLC), 32–33
Canadian National Railway (CN), 32
Canadian Pacific Railway (CPR), 155, 157
Canadian Westinghouse, 32
carburetors, 15–17
cast blocks, 57
Caterpillar Inc., 164
Centennial models, 130, 132
center-door configuration, 15
"centipedes," 80, 81, 169
Central of Georgia Railway (CG), 57, 90
Central Railroad of New Jersey (CNJ), 26, 30, 84, 92, 94, 99
Century of Progress Exposition, 34–37
Century series, 120, 122–23, 123, 128, 170
Champion, 49
Chatain, Henri, 16, 27
Chesapeake & Ohio Railway (C&O), 47, 67, 99, 111, 120
Cheshire, 42
Chicago, Illinois, 30, 34–37
Chicago, Milwaukee, St. Paul and Pacific Railroad (MILW): and
    Alco engines, 61–63; and Baldwin VO models, 57; and
    Fairbanks-Morse engines, 85–88, 86, 87; and road switchers,
    66, 91, 93, 102
Chicago, Rock Island and Pacific Railroad. *See* Rock Island Line
    (RI)
Chicago & North Western (C&NW): and AC propulsion, 153–
    54; and Alco engines, 61–63; and Baldwin engines, 81; and
    Fairbanks-Morse engines, 85; and gas turbine-electric locomo-
    tives, 127; and passenger locomotives, 46; and road switchers,
    94, 102; and SD series, 116
Chicago Burlington & Quincy (CB&Q): and the Century series,
    121; and diesel-hydraulic propulsion, 132; and EMC units, 50;
    and passenger locomotives, 44; and streamlined trains, 39–41;
    and wartime production, 58; and *Zephyrs*, 40, 41
Chicago Exhibition (1896), 6. *See also* Century of Progress
    Exposition
Chicago Great Western Railway (CGW), 22, 25
"chicken wire" units, 78
Chief passenger trains, 145
Chorlton, E. L., 32
circuit boards, 137
CIT Financial (CFEX), 164
City of Denver, 43
City of Los Angeles, 46
City of Salina, 40
City of San Francisco, 46
Clean Air Act, 159–60
Clement, Martin, 68
closed-loop control systems, 25
clutches, 15
coal trains, 146, 155–56, 158
Codrington, George, 21, 38, 45
Cohan, George S., 99
Coleman, L. G., 30
Columbus & Greenville Railway (CAGY), 93, 94
combustion, x, 1–3, 9–10, 24, 29, 159
Comet, 34
Commonwealth Steel, 32
commutators, 5, 5, 138
compression ignition, 27. *See also* diesel cycle

compression ratio, 9–10
compressors, 10–11, 126. *See also* superchargers
connecting rods, 9
Conrail (CR), 147
Consolidation Line, 88, 99
control systems: and AC propulsion, 150; and the Century series, 121; and diesel-hydraulic propulsion, 136; and EMC units, 23–25; and GE engines, 18; Lemp Control System, 24; and McKeens, 13–15; and microprocessors, 143; and solid-state electronics, 137; and wheel slip, 109
conversion of locomotives, 155–56
cooling systems, 89, 155
Cooper-Bessemer Corporation, 31, 60, 109, 135
Cotsworth, Marguerite, 40
Cotton Belt (SSW), 119
couplers, 49, 52, 86
cowcatchers, 13
crankshafts, 9, 17, 69, 121, 140
CSX Transportation (CSXT), 146, 155–57, 161
Cummins Inc., 31, 35
current, 6, 7–8. *See also* alternating current (AC) propulsion; direct current (DC) motors

Daimler, Gottlieb, 1
Daimler-Benz, 132
Dan Patch Lines (MN&S), 17, 18
Daniel Webster, 135
"dash 2" package, 138
Dash 8 series, 147, 148, 158, 171
Dash 9 series, 147, 149, 150, 155
Davidson, Robert, 5
Dayton Engineering Laboratories Company (Delco), 35
de Rochas, Alphonse Eugène Beau, 1
Defoe Shipbuilding Company, 35
Delaware, Lackawanna & Western (DL&W), 59–60, 99
Delaware & Hudson Railway (D&H): and Alco engines, 69–70, 75, 115, 122; and C628 units, 115; and GE engines, 17; and road switchers, 91–92; and Universal series, 115
demonstration tours, 40–41, 92–93, 111
Denver & Rio Grande Western (D&RGW): and AC power, 140; and Alco engines, 72; and "The Diesel Spirit" (Morgan), 173; and diesel-hydraulic propulsion, 134, 135, 136; and freight trains, 52; and "tunnel motors," 142
Denver Union Station, 41
Denver Zephyr, 44, 51
Depression, 33, 34
Detroit Diesel, 133
Dezendorf, Nelson, 102
Diesel, Rudolf: and Alco engines, 58; apprentices, 84; background, 2; and compression ignition, 27; and control systems, 18; and early diesel development, 37; and Winton engines, 21
diesel cycle, 4
Diesel Fireman vs. Western Association of Railway Executives, 44
"The Diesel Spirit" (Morgan), 173
diesel-battery hybrids, 164–67, 166
diesel-hydraulic propulsion, 131–36
Dilworth, Richard, 18, 23, 39, 44, 47, 101
diodes, 151, 152
direct current (DC) motors: and adhesion limits, 10; connection variations, 8; described, 5–7; and diesel-hydraulic propulsion, 131; and generators, 138; and McKeens, 16–17; and SD models, 149; and technological advances, 137
dirigibles, 32
displacement, 9, 10, 138–40
distillate engines, 16, 23, 39, 40
Donovan, M. J., 88
"doodlebugs," 22
Duetz MWM, 155
duplex drive, 67, 80–81, 84
dynamic braking, 52, 102, 146, 167
dynamometers, 111

E. & T. Fairbanks Company, 84
Eclipse Wind Energy Company, 84
economic pressures: and the Depression, 33–34; and dieselization, 67–68; and EMC, 23; and plant shutdowns, 143–47; and postwar production, 77–80; and recessions, 25; and the stock market crash, 25, 56; and strikes, 43–44, 70; and Westinghouse engines, 33–34. *See also* fuel consumption
Eddystone Works, 56, 93, 94
Edison, Thomas, 2, 6
Edward G. Budd Company, 39
Egbert, Percy, 72, 118–21
electric motors, 3, 5, 20. *See also* alternating current (AC) propulsion; direct current (DC) motors
Electro-Motive Company (EMC): and Brill cars, 20; and development funding, 37; and diesel evolution, 20–25; and freight trains, 51–53; and GE engines, 18; and GM, 37–38, 52; and nose styling, 46; and plants, 47–50, 61; and streamlined trains, 39, 41; and switchers, 47; and Winton Engine, 21, 47. *See also* General Motors Electro-Motive Division (EMD)
Electro-Motive Engineering Corporation (EMEC), 20–21
electronics, 137–38, 147, 160
Elgin Joliet & Eastern (EJ&E), 57, 84, 91
Elliot Manufacturing Company, 60
emissions, 159–60. *See also* pollution
Endsley, L. E., 85
engine configurations, 9
environmental issues, 159–60, 160–62, 164. *See also* pollution
Ephraim, Max, 101
ergonomics, 162
Erie Forge Company, 69
Erie Railroad (ERIE), 71, 72, 91–92, 109, 111
Evolution series, 158, 160
Executive Trains, 138
exhaust, 159–60. *See also* pollution

FA models (Alco), 68, 70, 70–73, 71
Fairbanks-Morse: and C-Liners, 87; and diesel advances, 169, 170; and GE engines, 27–28; and market share, 107, 108; and postwar production, 77, 84–88; and road switchers, 96–99, 97, 108; and switchers, 86
Faraday, Michael, 3
Farrington, J. D., 63
FB models (Alco), 70, 71, 72
FDL engines, 109, 118, 140
field coils, 5, 6, 24–25
flexibility, 80
Flying Yankee, 37, 41, 42
Flying Yankee Restoration Group, 41, 42

flywheels, *14*
Forney tank engines, 6
"44-toners," 63
four-stroke cycle engines, 2, 3, 10
FP7 models (EMD), 79
Freedom Train Operation, 73
freight units, 51–54, 70
Frisco (St. Louis–San Francisco Railway) (SLSF), 58, *64*, 111, *113*
FT models (EMD), 51–54, *53*, 78, *170*, *171*
fuel consumption: and dieselization, 68; and distillate engines, 16; and EMC, 23; and emissions, 167; and environmental issues, 160; and fuel tenders, 126; and gas turbine-electric locomotives, 125–28; and GMLG, 150; and Ingersoll-Rand engines, 28, 30

Galloway, Charles, 45
gas turbine-electric locomotives, *124*, 125–28, *126*, *127*
gas-electric cars, 20
Gaulard, Lucien, 7
gear ratios, 15, 63
GE-IR-Alco consortium, 58–63
Geitmann, F. J., 88
General Electric Appearance Design Division, 69
General Electric Company (GE): and AC propulsion, 140, 155–57, *156*, *157*, *158*; and Alco engines, 61, 72, *113*; and Baldwin, 55; and car body production, 85; and competition, 111, 171; and Dash 8 models, 148; and DC propulsion, 6–8, *149*; and diesel advances, 170; and early diesel development, 27–30; and EMC, 20; and emissions controls, 158, 160; and freight trains, 52; Gas Engine Division, 16–18, 27–28, 35, 63; and gas turbine-electric engines, *124*, 126; and hybrid engines, 167; and J. G. Brill, 20; and Jersey Central, 26; and joint marketing, 108; and leasing business, 143; Locomotive and Car Department, 63; and market share, 147; and numbering schemes, 138; and production levels, 118, *118*, 171–72, *172*; and road switchers, 106, 122; and second generation diesels, 63; and superchargers, *141*; and twin-diesel platforms, 128, 131; and the Universal series, 109, 111–18, *113*, *114*, 117, *174*; and WEMCo, 31–32
General Electric Locomotive and Car Department, 63
General Machinery Corporation, 88
General Motors Corporation: and competition, 111; and diesel advances, 169; and early diesel development, 34–37; and EMC acquisition, 52; and engine improvements, 143; and freight trains, 51–54; Locomotive Group, 150, 153–55, 171; and market share, 107, 147, 171; and opposed piston engines, 28; transfer of diesel operations, 143; and Winton motors, 35–39. *See also* General Motors Electro-Motive Division (EMD)
General Motors Electro-Motive Division (EMD): and AC propulsion, 138, 140, 153–55, 157; acquired by GE, 52; and Alco engines, 73, 122; and Baldwin engines, 81–84; and Blomberg trucks, 171; and Boise Locomotive, 164; and branch lines, 99–101, *100*; and competition, 111, 112–13, *113*, 171; and DC drive units, *149*; and diesel advances, 169–71; and "The Diesel Spirit" (Morgan), 173; and diesel-hydraulic propulsion, 136; and engine problems, 143; and environmental issues, 160, *161*; and expansion, 147; and experimental drives, 138; and export locomotives, 109; and Fairbanks-Morse engines, 88, 99; and 567 series engines, 69; and GM&O, *170*; and market share, 108; and passenger units, 48; and production levels, 69, 77–80, 118, *118*, 171–72, *172*; and road switchers, 91–92, 99–105, *100*, *101*, *103*, *105*, 107, 108, 122; and SD models, 116, 138, 140, 146, *174*; on

Sunset Route, *139*; transfer of diesel operations, 143; and twin-diesel platforms, 128, 129, *130*; and wartime production, 67
General Motors Locomotive Group (GMLG), 147, 160–62
General Pershing Zephyr, 50
General Steel Castings, 51, 57, 108
generators, 36, 167. *See also* hybrid motors
George Sheffield Company, 84
German Railway Office, 132
German State Railways, 132
Gibbs, John D., 7
GM-Siemens, 155
Goodyear, 34
GP models: and AC propulsion, 138, 140; color schemes, xii, *113*; cowling design, *112*; demonstrators, 147; development of, 101–105; evolution of, *139*; and freight trains, *170*; last models, 51; power increases, 113; and production levels, 170; and sales performance, 107; and tractive effort, *104*; and twin-diesels, 128; and unit reduction, 111–13
Gramme, Zènobe Théophile, 5
Grand Trunk Western Railroad (GTW), 66
Gray, Carl, 39
Great Lakes Steel, 34
Great Northern Railway (GN), 22, 52, 79
Green Diamond, 41, 43
Green Goats, 164–67, *166*
Green Kids, 166
Greenbriar Equity Group, 162
Gross Ton-Miles per Train Hour, ix
Grutzner, F. P., 84, 85
Gulf, Mobile & Northern Railroad, 60–61
Gulf, Mobile & Ohio (GM&O): and Alco engines, 61, 62, 63, 70–73; and Baldwin engines, 82; and competition, *170*; and EMC units, 21; and the GP series, 112; and PA models, 74; and road switchers, 65
Gulf Coast Rebel, 62, 74, 82
Gurley, Fred, 51–52, 57

Hale & Kilburn, 13
Hall Scott Engine Company, 18
Hamilton, Harold Lee, 20–22, 27, 37, 39–41, 45, 47
Harriman, Averell, 39
Harriman, Edward H., 2, 13, 34
"Heavy Duty," 96
Henry, Joseph, 3
Hershberger, David, 33
Hertz (measurement), 7
Heseltine, James, 18
high-speed trains, 39
highways, 23
horsepower, 9, 154–55, 160
hostile takeovers, 99
Hyatt Company, 16
hybrid motors, 8, 30, 164–67, *166*
hydraulic power transmission, 132
hydrodynamic braking, 136

I. P. Morris & De La Vergne Inc., 56–57
idlers, 11
Ignitrons, 150–51
"Ike," 32–33, *33*

Illinois Central Railroad (IC): C-C units, 31; and GE engines, 18; and the *Green Diamond*, 43; and rebuilding business, 162; and road switchers, 101; and streamlined trains, 41
Illinois Railway Museum, 102
Illustrated Bee (Omaha), 15
Imperial Engine Company, 28
Indiana Harbor Belt Railroad (IHB), 47, 96
Industrial Revolution, 1
Ingersoll, Simon, 28
Ingersoll Rock Drill Company, 2
Ingersoll-Rand: and Baldwin, 55; and diesels, 28–31; and early diesel development, 28–31; and Jersey Central, 26; and passenger locomotives, 44
Ingles, J. David, 170
injection (fuel), 36, 160
inline engines, 9
input frequency, 153
instrumentation, 24
intercoolers, 121
internal combustion (IC) engines, x, 1–3, 9–10, 24, 29, 159
International Motors Company, 20
inverters and inversion, 150–53, 155
Iranian Railway, 66
Irvin, William, 40
Issl, Max, 80

J. Armand Bambardier Ltd., 123
J. G. Brill Company, 18–20, 19
Jackson, Larry B., 85
Jay Street Connecting Railroad, 28, 59
Jersey Central. *See* Central Railroad of New Jersey (CNJ)
Junkers Company, 27
Justice Department, 54, 107

Kansas City, Ft. Scott & Memphis Railroad, 64
Kansas City Southern Railway (KCS): and the E series, 50; and environmental issues, 162; and Fairbanks-Morse engines, 86, 87; and hybrid engines, 167
Katy Railroad, 72, 146
Kaufman Act, 30, 55
Kettering, Charles, 35, 38, 40, 45, 50, 169
Kettering, Eugene, 35–36, 101
Knudsen Motor Company, 55
Korean War, 108
Kraus-Maffei Company, 134, 135, 136
Krupp engine, 56
Kuhler, Otto, 60, 61

labor unions, 43–44, 70, 80, 123, 143
Lackawanna Railroad. *See* Delaware, Lackawanna & Western (DL&W)
Langen, Eugen, 1
leasing, 143, 162
Lehigh & New England Railroad (LNE), 91
Lehigh Valley Railroad (LV): and Alco engines, 59, 72, 121; and Ingersoll-Rand engines, 30; and J. G. Brill, 20; and McIntosh & Seymour engines, 58
Lemp, Hermann, 18, 23, 27–29, 137
Lemp Control System, 23–25, 24
Lepitz, Alphonse, 132

Lima Locomotive Works, 52, 77, 88
Lima-Hamilton, 88–89, 89, 107
liquefied natural gas (LNG), 164, 165
liquid turbines, 132
LMX leasing, 143
Loewy, Raymond, 84, 85
London Midland and Scottish Railway, 32
Long Island Rail Road (LIRR), 92
Loop District, 30
Los Angeles Junction Railway (LAJ), 164, 165
Louisville & Nashville (L&N), 59, 71, 120

magnetism, 3–4, 5, 150
Maine Central Railroad (MEC), 41, 42
Manhattan, New York, 6, 30
Manhattan Railway, 6
marine engines, 21, 38, 52, 56–57, 88
Mark Twain Zephyr, 41
market share, 108, 122, 122, 147, 157, 171
Massachusetts Institute of Technology (MIT), 40
materials issues, 36
Maybach, Wilhelm, 1
Maybach Company, 132, 135
McCook, Illinois, 47
McIntosh, John E., 58
McIntosh & Seymour, 58–59
McKeen, William R., 13, 34
McKeen Motor Car Company, 12, 13–16, 14, 16, 20, 25
Mean Effective Pressure (MEP), 9–10
Meissner, Edwin E., 21
Mekydro transmissions, 132, 135
Menlo Park, New Jersey, 6
mergers, 52, 89, 99, 143, 148
Mexican National Railway, 75, 81, 109
Mexico, 92, 96
Michelin, 39
microprocessors, 143, 147
Mid-Continent Railway History Museum, 22–23
"Mike," 32–33, 33
military equipment, 108
Milwaukee Coke and Gas Company, 31
Milwaukee Road. *See* Chicago, Milwaukee, St. Paul and Pacific Railroad (MILW)
Minneapolis & St. Louis Railway (M&StL), 54, 68
Minneapolis-Rochester Line, 25
Miss Lou, 43
Missouri Pacific Railroad (MP): acquired by UP, 148; and Alco engines, 72, 75; and Baldwin engines, 81–84; and GP models, 139; and road switchers, 104; Taylor, Texas yard, 174; and the *Texas Eagle*, 74
Missouri-Kansas-Texas Railroad (MKT), 127, 148
mixed diesel cycle, 3
M-K Rail Corporation, 164
modular controls, 137
modular locomotives, 80
Monon Railroad (MON), 76, 88, 100
monopolies, 54, 107–23
Montana RailLink (MRL), 162
Montreal Locomotive Works, 108, 123
"More Power to America," 73

Morgan, David P., 33, 41, 105, 135, 172–73
Morrison-Knudsen Co., 162–64, *163*
Morse, Charles H., 84
MotivePower International (MPI), 164
multiple cylinder engines, 9, 20
Museum of Science and Industry (Chicago), 41

Nashville Chattanooga & St. Louis Railway (NC&StL), 61, *102*
National Cash Register, 35
National Railway Equipment, 167
National Railways of Mexico, 75, *81*, 109
Neuhart, D. S., 128
New Haven. *See* New York, New Haven and Hartford Railroad (NH)
New Orleans Public Belt Railroad (NOPB), 56, 57
New York, New Haven and Hartford Railroad (NH): and Alco engines, 59, 63, 70, 72; and Baldwin engines, 94; and the Century series, 120; and diesel-hydraulic propulsion, 133, 135; and J. G. Brill, 20; and road switchers, 91–92; and Westinghouse engines, 34
New York, Ontario & Western Railway (NYO&W), 54
New York, Susquehanna & Western (NYS&W), 66
New York Central Railroad (NYC): and Alco engines, 58, 59, 72, 122; and Baldwin engines, 82, 83, 84; and diesel-hydraulic propulsion, 133, 135; and road switchers, 66, 91–92, 96, *103*; and the Universal series, 111
New York City, 29
Nickel Plate (NKP), 72, 75, 88, *112*, 126
nicknames for locomotives: alligators, *119*; battleships, 15; centipedes, 80, *81*, *169*; chicken wire units, 78; doodlebugs, 22; "44-toners," 63; Ike & Mike, 32–33, *33*; modern translations, 172; potato bugs, 15; U-Boats, 111
noise pollution, 30, 159
Norfolk & Western Railway (NW), 67, 110
Norfolk Southern (NS), 64, 94, 95, 167
North Africa, 94
North American Cab, 143
Northampton & Bath, 33
Northern Pacific Railway (NP), 51, 58, *119*, *121*
nose styling, *46. See also* shovelnose designs
numbering schemes, 43–44, 46, 80, 101–102, 111, 122, 138

Oakway Leasing, 143
Ohm's Law, 6
oil engines, 28–29
oil refining, 23
Olympian Hiawatha, 85–88
opposed-piston (O-P) engines: and the Consolidation line, 88; described, 9; and Fairbanks-Morse engines, 96, 99; and GE, 17, 27–28; and submarine diesels, 85; and Train Masters, 98
Oregon & California Railroad, 12
Oregon Railway and Navigation Company, 15
Osborne, Cyrus, 101
Otto, Nikolaus August, 1, 21
Otto engines, 2–3, *3*

Pacific Electric Railway (PE), 15
Pacific Harbor Line (PHL), 164–67
Pacinotti, Antonio, 5
Page, Charles G., 5

pancake configuration, 9
parallel configurations, 6–7, 8, 17, 28
Paris Exhibition (1867), 1
passenger units, 13, 44–47, *48*, 70, 72, 133
Paterson, V. H., 88
Patten, Ray, 91
Patten, Richard, 69
Penn-Central, 111
Pennsylvania Railroad (PRR): and AC propulsion, 150–51; and Alco engines, 72, 122; and Baldwin engines, 58, 81, *81*, 83, 84; and dieselization, 67–68; and Fairbanks-Morse engines, 86; and gas turbine-electric locomotives, 126–27; and GE 44-toners, 63; and J. G. Brill, 20; and Lima-Hamilton units, 89; and road switchers, 96, *103*; and steam locomotives, 67; and switchers, 68; and the Universal series, 111; and Westinghouse engines, 34
Phelps-Dodge, 102–104
Pittsburgh & Lake Erie Railroad (P&LE), 72
Pittsburgh & West Virginia Railway (PWV), 96
pneumatic engine controls, 136–37
polarity, 3, 150. *See also* alternating current (AC) propulsion
pollution, 30, 55, 159–60, 164–67
Ponce de Leon, 62
Port Terminal Railroad, *163*
porthole windows, 14, 80
Portland, Oregon, 15
"potato bugs" (M-1s), 15
Power River Basin, 147
power transmission, 137
Precision National Corporation, 162
pressure-volume (P-V) diagrams, 2–3, *3*, *4*
Price, William T., 28
Price Combustion Chamber, 29
Priest, S. D., 16
production levels: and Alco engines, 69; and EMD, 118, *118*, 170–72, *172*; and freight trains, 54; and GE, 118, *118*, 172; road switchers, 122; SD series, 116; and the War Production Board (WPB), 52, 63, 66–67, 77–80, 85
publicity, 40–41, 92–93, 111, 135
Pullman Company, 22, 33, 41, 51
Pullman-Standard Car Manufacturing Company, 39, 43, 133

radial trucks, 147–48
radiators, 109
Rail Diesel Cars (RDC), 132–33, *133*
RailPower Technologies Corporation, 164–66, 167
Railserve, *166*
Rand Drill Company, 28
Rathbun, George, 29
RC buffer circuits, 150
Reading Railroad (RDG), 20, 57, 92, 99
Rebels, 61, 62
rebuilding locomotives, 122, 162
rectification, 138, 150–53, *151*
reliability, 109, 133, 137, 138–40
remote control, 167
Rentscher, George, 95
resistance (electrical), 6
restoration efforts, 41, 75
revolutions per minute (rpm), *4*, 9–10, 125–26, 143

rheostats, 23
Richmond Union Passenger Railway, 6
Rio Grande Railroad. *See* Denver & Rio Grande Western (D&RGW)
road switchers, 95–105; and Alco engines, 63–66, *65*, 91–92, *100*, *108*, *119*, 121, 122, 170; and American Locomotive, 91–93; and Baldwin engines, 93–95, *94*; and branch line models, *100*; and EMD, 99–105, 107; and Fairbanks-Morse engines, 97; and GE, 109–11; and market share, *108*; and Norfolk Southern, 95; production levels, 122; Seaboard Air Line, *92*, 96; Southern Pacific, 93, 96; Train Masters, 98
Rock Island Line (RI): and Alco engines, 61–63; and Baldwin engines, 81–84; and bi-power units, 30; and diesel-hydraulic propulsion, 133; and EMC engines, 23; and GP models, *139*; and McKeen retrofits, 25; and road switchers, 66, 92, 99; and *Rockets*, 50, *51*; and the Universal series, 111
Rockets, 50, *51*
Royal Blue, 44
rubbing contacts (brushes), 5

Sahara Desert, 94
Salisbury, Carl, 35–36
Santa Fe Railroad. *See* Atchison Topeka & Santa Fe (AT&SF)
Savannah & Atlanta Railroad, 94
Schmidt, Henry, 88
Schneider, Heinrich, 85
Scooter, 19
SD series locomotives: and AC propulsion, 140, 153, *154*, 154–55, 171; and Alco engines, 122; and C-C units, *114*; and competition, 111–13; and DC propulsion, *149*; and demonstrators, *148*; and design improvements, *146*, 169; and diesel-hydraulic propulsion, 136; and environmental issues, 160; and Evolution series, *161*; and leasing, 143; popularity of, 170–71; and power limits, 118; and production levels, 116; and sales performance, 140, 147–50; and tractive effort, *104*; and "tunnel motors," *142*; UP's acquisition of, 148–50
Seaboard Air Line Railroad (SAL): and Alco Century series, 122; and Baldwin engines, 80–81, 84; and EMC units, 50; and road switchers, *92*, 92, 96; and streamlined trains, 43
Seaboard Coast Line Railroad (SCL), 117, *120*
self-propelled coaches, 13–25
semiconductors, 151
Semmering Incline, 135
Senate Subcommittee on Antitrust and Monopoly, 54, 107
Sergeant Drill Company, 28
series configuration, 6–7, *8*, 17
Service Motors Company, 19, 20
Seymour, James L., 58
Sharknose units, 75, 83, *84*, *127*, 170
Sherman Hill, 40
Shoemacher, F. G., 36
shot welding, 39
shovelnose designs, 40, 43, 44, 61
shunt motors, 6–7
Siemens, Ernst Werner von, 6
Siemens Transportation Group, 153
silicon, 151
size of locomotives, 80
slippage, 11. *See also* adhesion limits
Sloan, Alfred P., 16, 37, 40

Society of Automotive Engineers, 35
solid-state electronics, 137, 147, 160
Soo Line Railroad (SOO), 94, 102
Southern Pacific Railroad (SP): and AC propulsion, 140, *156*; acquisition by UP, 148; and Alco engines, 72, 75; and Baldwin engines, 58; and diesel-hydraulic propulsion, 134, 135, 136; and FT models, 54; and GP models, *144*; and J. G. Brill, 20; and locomotive innovations, 125; and McKeens, 12, *14*, 15; and passenger trains, 144–45; and rebuilding business, 162; and road switchers, 93, 96, 99, *103*, 119; and the SD series, *116*; and Sunset Route, *139*; and "tunnel motors," *142*; and twin-diesel platforms, 128; and the Universal series, 111
Southern Railway (SOU): and Alco engines, 61–63, 122; and Fairbanks-Morse engines, 85; and FP7 units, *79*; and freight trains, 51, 52; and road switchers, 91; and supercharged engines, *141*; and the Universal series, 110
Southerner, 79
spalling, 36
spark ignition, 2, 3, 17, 28
special duty (SD) units, 105
Sperry Company, 35
Spokane Portland & Seattle Railway (SPS), *121*
Sprague, Frank J., 6
Sprague Electric, 20
St. Louis Car Company, 33, 43, 85
St. Louis Exhibition (1904), 2
St. Louis Southwestern Railway (Cotton Belt) (SSW), *119*
Standard Motor Construction Company, 13–15
Stanley, John, 102
steelworkers, 70
steeple cap design, 6
stock market crash, 25, 56
Stotz, John K., 85
strikes, 43–44, 70
Studebaker-Worthington, 123
Sulzer Brothers, 60
Super Chief, 46, 143
Super Power design, 67
Super Steel Schenectady, 147
superchargers: and AC propulsion, 155; and Alco engines, 60, 69, 121; background of, 10–11; and Baldwin engines, 80, 81; and emissions, 160; and GE engines, *141*; and road switchers, 105; and technological advances, 137, 138–40, 169
switchers: and Alco engines, 60; and Baldwin engines, 57; and diesel-battery hybrids, 164–67, *166*; and EMC engines, 47; and EMD engines, *104*, 170; and Fairbanks-Morse engines, 85, 86; and the Jersey Central, 26; and Lima-Hamilton engines, 89; and passenger locomotives, 47; and Pennsy, 68; and Westinghouse engines, 34, *34*. *See also* road switchers
Swope, Gerard, 40
Symes, James M., 68
synchronization, 35

Tennessee Alabama & Georgia Railway (TAG), 19
Tennessee Pass, 173
Texas & New Orleans Railroad (TNO), 103
Texas Eagle, 74
Texas Emissions Reduction Program (TERP), 167
Texas Rocket, 50
Texas Zephyr, 49

Thoroughbred, 76

Three Rivers Electric Company, 84

three-phase AC, 152

thyristers, 151, 152

Tigrett, Isacc, 60–61, 68, 70–72

Toledo Peoria & Western Railway (TPW), 91

top dead center (TDC), 2, 160

torque, 153

torque converters, 132

traction motors: and AC power, 153; and control systems, 23–25, 24; DC connection variations, 8; and dynamic brakes, 52; and GE engines, 16–17; and gear ratios, 63; and technological advances, 137; and tractive effort, 10, 11, 156–57

Train Masters, 98, 99, 122, 169

Train X, 133

Trains (magazine), 41, 135, 170, 172, 173

transformers, 7–8

transmission lines, 7–8, 13–15

Trieber Company, 35

tri-power units, 31

trucks: Blomberg trucks, 117, 171; Blunt-trucks, 59; and improvements, 92; radial trucks, 147–48, 150; and *Super Chief*, 46

"tunnel motors," 142

turbochargers: and Alco engines, 60, 69, 121; and Baldwin engines, 80; and diesel-hydraulic propulsion, 135; and efficiency, 93; and Lima-Hamilton engines, 88; and replacement of old diesels, 112; and road switchers, 105; and technological advances, 138–40; and turbine combustion, 125–26

Twin Zephyrs, 41

twin-diesel platforms, 128–31, 129, 130, 131, 132

two-stroke cycle engines, 2, 9, 36

unemployment, 68

Union Pacific Railroad (UP): and AC propulsion, 153–55, 157; and Alco engines, 72; and the *City of Salina*, 40; and Dash 8 models, 148; and DC propulsion, 149; and Electro-Motive units, 46, 46; and environmental issues, 160, 162; and Fairbanks-Morse engines, 85; and FT models, 54; and gas turbine-electrics, 124, 126, 126–27; and hybrids, 164, 167; and J. G. Brill, 20; and LNG units, 164, 165; and locomotive innovations, 125; and marketing strategies, 148; and McKeen engines, 15, 16, 25; and numbering schemes, 43–44; and Pullman Standard trains, 43; and repair and maintenance issues, 148–50; and road switchers, 96; and SD series, 148–50, 154; and streamlined trains, 39; and supercharged engines, 141; and twin-diesel platforms, 128, 129, 130; and the Universal series, 109, 111; and Winton engines, 40–41

Union Railroad, 91

unions, 43–44, 70, 80, 123, 143

Universal series: and AC propulsion, 138; and C-C units, 114, 115; and competition, 112–13, 113; and cooling modules, 131; and diesel advances, 169; and GE, 110; and Missouri Pacific, 174; popularity of, 170; and power limits, 118; and the road-switcher market, 109–11; and sales performance, 147; and the Seaboard Coast Line, 117; and superchargers, 141; and twin-diesel platforms, 128, 129

U.S. Department of War, 66

U.S. Environmental Protection Agency, 159–60

U.S. Navy, 36

U.S. Railroad Administration, 20, 54

U.S. Senate, 27

U.S. Steel, 66

Utah Railway (UTAH), 164

valve control, 9

Vauclain, Sam, 55

Vaughan, Paul, 118–21

Vee-pattern engines: and Alco engines, 69; described, 9; and GE, 17; and McIntosh & Seymour engines, 58; and Westinghouse engines, 32

VIA Rail, 133

Virginia & Truckee Railroad (VT), 16

Virginian Railway (VGN), 67, 99

visibility, 34, 104

Voith transmissions, 132, 135–36

voltage, 6–8

Wabash Railroad (WAB), 72, 111

Wabtec Corporation, 164

Waldorf-Astoria Hotel, 72

War Production Board (WPB), 52, 63, 66–67, 77, 80, 85

Ward-Leonard control system, 24

Wason Manufacturing Company, 17

wave-chopping, 151, 152

weight of locomotives, 68, 136

West Side Line, 29, 30

Western Pacific Railroad (WP), 113, 121, 133, 139, 148

Westinghouse, George, 7, 8

Westinghouse Electric and Manufacturing Company (WEMCo): and AC propulsion, 8; and Baldwin engines, 55; and Baldwin-Lima-Hamilton, 95; and diesel development, 31–34; and gas turbine-electrics, 126–27, 127; and market share, 107–108; and merger negotiations, 89; and switchers, 33, 34

wheel slip control, 11, 111, 138, 143, 150

White Motors, 20, 22

Whyte system, 11

Wiffenbach, John, 88

William Beardmore Co., 32

wind tunnel testing, 40

Winton Engine Company: and the Chicago Exhibition, 40; and EMC units, 21, 21, 22; and gas-electrics, 23; and General Motors, 35–37, 37–38, 39; and streamlined trains, 43–44; and switchers, 47; and the 201A engine, 37, 47

Woolsey Tool & Motor Car Company, 16–17

World War I, 16, 18

World War II, 67, 169

World's Colombian Exhibition (1896), 6

Worthington Corporation, 123

Xplorer, 135

Zephyrs: on the CB&Q, 41, 132; christening, 40; *Denver Zephyr*, 44, 51; development of, 40, 51–52, 169; *General Pershing Zephyr*, 50; *Mark Twain Zephyr*, 41; success of, 104; testing, 40–41; *Texas Zephyr*, 49; *Twin Zephyrs*, 41; *Zephyrette*, 133

After a 42-year career as a practicing engineer and university faculty member, **J. Parker Lamb** retired in 2001 to pursue his lifelong interest in the history of American technology, with an emphasis on railroad developments. Beginning in 1991, he has authored four books, including *Perfecting the American Steam Locomotive* (Indiana University Press, 2003), and has contributed hundreds of photographs to magazines and books since 1954. He lives in Austin, Texas.

BOOKS IN THE RAILROADS PAST AND PRESENT SERIES:

Landmarks on the Iron Railroad: Two Centuries of North American Railroad Engineering by William D. Middleton

South Shore: The Last Interurban (revised second edition) by William D. Middleton

"Yet there isn't a train I wouldn't take": Railroad Journeys by William D. Middleton

The Pennsylvania Railroad in Indiana by William J. Watt

In the Traces: Railroad Paintings of Ted Rose by Ted Rose

A Sampling of Penn Central: Southern Region on Display by Jerry Taylor

The Lake Shore Electric Railway by Herbert H. Harwood, Jr. and Robert S. Korach

The Pennsylvania Railroad at Bay: William Riley McKeen and the Terre Haute and Indianapolis Railroad by Richard T. Wallis

The Bridge at Quebec by William D. Middleton

History of the J. G. Brill Company by Debra Brill

When the Steam Railroads Electrified by William D. Middleton

Uncle Sam's Locomotives: The USRA and the Nation's Railroads by Eugene L. Huddleston

Metropolitan Railways: Rapid Transit in America by William D. Middleton

Limiteds, Locals, and Expresses in Indiana, 1838–1971 by Craig Sanders

Perfecting the American Steam Locomotive by J. Parker Lamb

From Small Town to Downtown: A History of the Jewett Car Company, 1893–1919 by Lawrence A. Brough and James H. Graebner

Steel Trails of Hawkeyeland: Iowa's Railroad Experience by Don L. Hofsommer

Still Standing: A Century of Urban Train Station Design by Christopher Brown

The Indiana Rail Road Company: America's New Regional Railroad by Christopher Rund

Amtrak in the Heartland by Craig Sanders

The Men Who Loved Trains: The Story of Men Who Battled Greed to Save an Ailing Industry by Rush Loving Jr.